Frontiers in Mathematics

Advisory Editors

Laurent Saloff-Coste, Cornell University, Ithaca, NY, USA
Igor Shparlinski, The University of New South Wales, Sydney, NSW, Australia
Wolfgang Sprößig, TU Bergakademie Freiberg, Freiberg, Germany

This series is designed to be a repository for up-to-date research results which have been prepared for a wider audience. Graduates and postgraduates as well as scientists will benefit from the latest developments at the research frontiers in mathematics and at the "frontiers" between mathematics and other fields like computer science, physics, biology, economics, finance, etc. All volumes are online available at SpringerLink.

Ludmila Nikolova • Lars-Erik Persson • Sanja Varošanec

Continuous Versions of Some Classical Inequalities

Ludmila Nikolova
Faculty of Mathematics and Informatics
Sofia University "St. Kliment Ohridski"
Sofia, Bulgaria

Sanja Varošanec
Department of Mathematics, Faculty of Science
University of Zagreb
Zagreb, Croatia

Lars-Erik Persson
Department of Computer Science and
Computational Engineering
UiT - The Arctic University of Norway
Narvik, Norway

Department of Mathematics
Uppsala University
Uppsala, Sweden

ISSN 1660-8046 ISSN 1660-8054 (electronic)
Frontiers in Mathematics
ISBN 978-3-031-83371-7 ISBN 978-3-031-83372-4 (eBook)
https://doi.org/10.1007/978-3-031-83372-4

Mathematics Subject Classification: 26D10, 25D15, 39B62, 46E70, 25D20, 46E27

© The Editor(s) (if applicable) and The Author(s), under exclusive license to Springer Nature Switzerland AG 2025

This work is subject to copyright. All rights are solely and exclusively licensed by the Publisher, whether the whole or part of the material is concerned, specifically the rights of translation, reprinting, reuse of illustrations, recitation, broadcasting, reproduction on microfilms or in any other physical way, and transmission or information storage and retrieval, electronic adaptation, computer software, or by similar or dissimilar methodology now known or hereafter developed.
The use of general descriptive names, registered names, trademarks, service marks, etc. in this publication does not imply, even in the absence of a specific statement, that such names are exempt from the relevant protective laws and regulations and therefore free for general use.
The publisher, the authors and the editors are safe to assume that the advice and information in this book are believed to be true and accurate at the date of publication. Neither the publisher nor the authors or the editors give a warranty, expressed or implied, with respect to the material contained herein or for any errors or omissions that may have been made. The publisher remains neutral with regard to jurisdictional claims in published maps and institutional affiliations.

This book is published under the imprint Birkhäuser, www.birkhauser-science.com by the registered company Springer Nature Switzerland AG
The registered company address is: Gewerbestrasse 11, 6330 Cham, Switzerland

If disposing of this product, please recycle the paper.

Photo by Dr. Hana Turčinová. Northern light in Luleå, Sweden—a common appearance during the writing of this book near *Hotel Infinity*

Preface

Inequalities is an extremely important research area in analysis and the interest has even increased very much especially during the last years. One of the reasons for this is the fact that this is a very fascinating and creative research area in itself but also, and most essentially, because of its importance for the development of other areas of mathematics, such as Fourier Analysis, Interpolation Theory, Approximation Theory, Harmonic Analysis, Function Spaces, and even Engineering Mathematics and its applications. In particular, today there are even more than 50 monographs and several journals completely devoted to this subject. A more complete description of this background and literature is presented in the beginning of our Chap. 1.

It is more recently discovered that parts of this theory of inequalities can be generalized and given in a more general continuous/family form. In order to provide our readers with the first illustrative example of this idea we consider the classical Hölder inequality in the following form (i.e., as an inequality between two geometric means): Let p and q be positive numbers such that $p > 1$, $1/p + 1/q = 1$ and f_0 and f_1 be non-negative measurable functions on the measure space (Y, ν). Then

$$\int_Y f_0^{1/p}(t) f_1^{1/q}(t) \, d\nu(t) \leq \left(\int_Y f_0(t) \, d\nu(t) \right)^{1/p} \left(\int_Y f_1(t) \, d\nu(t) \right)^{1/q}. \tag{1}$$

Hence, in this situation we have two functions f_0 and f_1 involved. There are also versions with n, $n = 3, 4, \ldots$ functions involved. In our continuous/family situation we replace $f_0(t)$ and $f_1(t)$ with $f_s(t) = f(s, t)$ which is a family of s functions involved, where s belongs to, for example, $\mathbb{R}$ or the unit circle in the complex plane (the last example is used when we explain a connection with interpolation theory between families of Banach spaces). Also here we can prove a Hölder-type inequality now with some generalized forms of geometric means involved. We also remark that if we let the function f_s only take two "values" f_0 and f_1 we get back our original Hölder inequality. This is explained in detail in Remark 1.2 in Chap. 1 and later also proved.

In this book we state, prove, and discuss a number of classical inequalities in such continuous/family forms. In particular, we prove such more general versions of the following inequalities: Hölder's inequality, Minkowski's inequality, Beckenbach-Dresher's inequality, Jensen's inequality, Jensen-Mercer's inequality, Popoviciu's inequality, Bellman's inequality, and Hardy's inequality. Moreover, the corresponding reversed inequalities in continuous/family forms are given. It is also well-known that many of the classical inequalities hold also in a refined form, usually by adding some positive term to the smaller left-hand side. Indeed, we also state and prove that also such refinements can be proved in the more general continuous/family frame we are working in.

The first time the present authors met the need to develop inequalities (mostly of Hölder-type) in such continuous/family forms was in connection to that they were involved in the development of creating an interpolation theory between a family of/continuously many Banach spaces. Note that classical interpolation theory concerns usually interpolation between only two (or at most finitely many) Banach spaces. There are many similarities between these theories, e.g., the one pointed out above: to generalize inequalities for (interpolation between) two functions (Banach spaces) to cover a full continuous/family situation. These thoughts have influenced the text on some places and especially of Appendix, where we have given our readers a very short introduction how interpolation between families of Banach spaces can be understood for a more broad audience.

Another guiding idea for us is convexity. It is known that both convexity (almost equivalent to Jensen's inequality) and interpolation are very powerful tools to prove inequalities. And, more important, there are close connections between these areas. And these relations are also partly kept in our continuous/family situation. Also, this fact has influenced our way of thinking and presenting in this book, see especially Appendix.

Summing up: This book is the first one focused mainly on inequalities of these continuous/family forms. We hope that this interplay between classical theory of inequalities and these newer continuous/family forms, including some corresponding open questions, will be very useful for a broad audience of readers and serve as a source of inspiration for further research in this fascination area. Moreover, we also hope that new applications can be pointed out outside those we have described concerning Interpolation between families of Banach spaces and Hardy-type inequalities.

Acknowledgments

Author Ludmila Nikolova would like to thank Faculty of Mathematics and Informatics of Sofia University "St. Kl. Ohridski" and also other two authors' universities in Sweden and Croatia, for the support and corresponding travel grants, which provided numerous opportunities for us three co-authors to meet and have this fruitful collaboration, including the writing of several joint papers and furthermore this book. Author Lars-Erik Persson thanks (the Swedish government supported) International Science Program (ISP) in Uppsala for financial support, which has been crucial to do some of the research equipped with this book. Author Sanja Varošanec thanks the Department of Mathematics of Faculty of Science in Zagreb for allowing her to use the sabbatical year during which this book was completed.

We are very grateful to several researchers around the world (co-authors, colleagues, etc.) for various kinds of contributions and support, which essentially have contributed to improve the quality of the content of this book. As typical examples in this "supporting team" we want to mention Sorina Barza, María J. Carro, Alois Kufner, Constantin Niculescu, Josip Pečarić, Natasha Samko, and Theodossios Zachariades.

We also thank Hana Turčinová for permitting us to use her photo of the magic Nordic light we have chosen to use on the frontpage as a symbol of this book. It is taken exactly at the place ("Hotel Infinity") where the second-named author took a part of writing this book.

We sincerely thank all three reviewers for their very useful evaluations and suggestions, which improved the presentation and content of the final version of this book.

Finally, and most important, our most cordial thanks go to our wonderful families for their patience, encouragement, support, and love during all our work with this book.

Competing Interests The authors have no competing interests to declare that are relevant to the content of this manuscript.

Sofia, Bulgaria	Ludmila Nikolova
Uppsala, Sweden	Lars-Erik Persson
Zagreb, Croatia	Sanja Varošanec
November 2024	

Contents

1 **Continuous Forms of Classical Inequalities** 1
   1.1 Introduction 1
   1.2 Continuous Forms of the Hölder and Minkowski Inequalities 5
       1.2.1 On the Hölder Inequality 5
       1.2.2 On the Minkowski Inequality 8
       1.2.3 Relation Between the Hölder and Minkowski Inequalities 11
       1.2.4 On the Beckenbach-Dresher Inequality 13
   1.3 Continuous Forms of the Popoviciu and Bellman Inequalities 15
       1.3.1 On the Popoviciu Inequality 15
       1.3.2 On the Bellman Inequality 20
   1.4 On the Gauss-Pólya Inequality 24

2 **Refinements of Continuous Forms of Inequalities** 31
   2.1 Refinements of the Hölder and Popoviciu Inequalities 32
   2.2 Refinements of the Minkowski and Bellman Inequalities 39
   2.3 Refinements of the Jensen Inequality and Its Reversed Version ... 43
   2.4 Refinement of the Jensen-Mercer Inequality 47
   2.5 Refinements of the Hardy Inequality 50

3 **Refinements of Inequalities via Strong Convexity and Superquadracity** ... 57
   3.1 On the Jensen Inequality for Strongly Convex Functions 57
   3.2 Refinements of Some Hermite-Hadamard-Type Inequalities for Strongly Convex Functions 63
   3.3 Refinements of the Jensen Inequality via Superquadracity 65
   3.4 Refinements of the Continuous Minkowski and Beckenbach-Dresher Inequalities via Superquadracity 69
   3.5 Refinements of the Continuous Hardy Inequality 75

4 **Functionals Associated with Continuous Forms of Inequalities** 79
   4.1 Properties of Some Hölder-Type Functionals 79
   4.2 Properties of Some Minkowski-Type Functionals 85
   4.3 Properties of Some Jensen-Type Functionals 88
   4.4 Properties of Some Gauss-Pólya Functionals 93

5 Some Classical Inequalities Involving Banach Lattice Norms 99
    5.1   Hölder- and Minkowski-Type Inequalities 100
    5.2   Popoviciu- and Bellman-Type Inequalities 107
    5.3   Beckenbach-Dresher-Type Inequalities 113
    5.4   Some New Hardy-Type Inequalities in Banach Function Spaces 118

A Appendix 127

References 137

Index 143

Continuous Forms of Classical Inequalities 1

1.1 Introduction

This section is written mainly as an introduction for this chapter, but it also contains information, facts, and ideas, which partly can be regarded as an introduction to all parts of the book.

Any mathematician will agree that inequalities play an important role as in independent research area. Today we are witnessing of a very strong development of the theory of inequalities. Currently, many mathematicians are working with these problems and many articles are published. Due to the large number of individual articles and contributions, books appear from time to time that try to capture as much of this knowledge about inequalities and that give us an insight into the entire field. Today there are already published more than 50 such monographs. Here we will highlight some of these general monographs, such as, for example: [14, 47, 80, 82, 106]. Besides these general monographs, there are books that deal with some special types of inequalities, such as [6, 11, 20, 21, 25, 39, 43, 55, 56, 65, 67, 69, 72, 105], where it is evident from the titles which narrower types of inequalities are considered.

We claim that this great and increasing interest in this research field is not only that it is fascinating in itself but also for its great importance for the development of some related research fields, such as for example: Convexity Theory, Fourier Analysis, Interpolation Theory, Approximation Theory, Function Spaces, Harmonic Analysis, Engineering Mathematics, Theory of Means, etc. Therefore, also in books that deal with various aspects of these fields, we can find a lot of inequalities, mostly as applications of known inequalities but also proofs of new inequalities, which were important to have for the further development of the actual field. Just as typical examples of such books, we recommend the following to the interested reader: [23–25, 59, 77, 84, 85, 110, 122].

Among the numerous inequalities, some stand out, which we call classical inequalities. These are inequalities that are usually associated with the names of some well-known mathematicians, who first discovered and/or proved them, such as: Jensen, Gauss, Cauchy, Hölder, Minkowski, Popoviciu, Bellman, Chebyshev, Lyapunov, Bessel, Hadamard, Hilbert, Hardy, Pólya, Markoff, Beckenbach, Dresher, Young, Hausdorff, Hermite, etc.

In this book, we will especially consider some of these classical inequalities, e.g., Hölder's, Minkowski's, Jensen's, their reverse versions so-called Popoviciu's and Bellman's, Beckenbach-Dresher's, and Hardy's. The reason for this is that in all these cases, this book contains generalizations of them to the continuous/family forms, which is the main focus of this book. Since this inequality book is the first one devoted on this subject, we raise the following open question, which we think can be of great interest for the inequality community described above:

Open Question To generalize also other known inequalities in this continuous/family direction.

It is also very challenging to find new applications outside those we have pointed out in this book, i.e., those in Interpolation Theory and Hardy-Type Inequalities. To give our readers a quick introduction to this chapter and, more generally, the meaning of an inequality in continuous form, we concentrate on the classical Hölder inequality.

Example 1.1 The discrete Hölder inequality reads: *If $a_1, \ldots a_n, b_1 \ldots, b_n$ are nonnegative numbers and if p and q are positive numbers such that $\frac{1}{p} + \frac{1}{q} = 1$, then the following inequality holds:*

$$\sum_{i=1}^{n} a_i b_i \leq \left(\sum_{i=1}^{n} a_i^p\right)^{1/p} \left(\sum_{i=1}^{n} b_i^q\right)^{1/q}. \tag{1.1}$$

The integral Hölder inequality reads: *If p and q are positive numbers such that $\frac{1}{p} + \frac{1}{q} = 1$ and if f and g are nonnegative functions on the measure space (Y, ν), such that $f \in L_p(Y)$, $g \in L_q(Y)$, then*

$$\int_Y f(t)g(t)\,d\nu(t) \leq \left(\int_Y f^p(t)\,d\nu(t)\right)^{1/p} \left(\int_Y g^q(t)\,d\nu(t)\right)^{1/q}. \tag{1.2}$$

A continuous version of (1.1) and (1.2) reads (see Theorem 1.1):

Let u and v be weight functions on X and Y, respectively, such that $\int_X u(x)\,d\mu(x) = 1$. Let f be a positive function on $X \times Y$ and measurable with respect to a measure $\mu \times \nu$. Then

1.1 Introduction

$$\int_Y \exp\left(\int_X \log f(x,y) u(x) \, d\mu(x)\right) v(y) \, d\nu(y)$$
$$\leq \exp\left(\int_X \log\left(\int_Y f(x,y) v(y) \, d\nu(y)\right) u(x) \, d\mu(x)\right). \quad (1.3)$$

Remark 1.1 Let us consider a particular case. Namely, let $u(x) = 1$, $X = X_1 \cup X_2$, $X_1 \cap X_2 = \emptyset$ with $\int_{X_1} d\mu(x) = \frac{1}{p}$, $\int_{X_2} d\mu(x) = \frac{1}{q}$, where $\frac{1}{p} + \frac{1}{q} = 1$ and

$$f(x,y) = \begin{cases} f^p(y), & x \in X_1 \\ g^q(y), & x \in X_2. \end{cases}$$

Then (1.3) collapses to the integral Hölder inequality (1.2). Similarly, using a discrete measure on Y, we get the discrete Hölder inequality (1.1).

Hence, inequality (1.3) can be considered as a generalization of inequalities (1.1) and (1.2), and we call it the continuous form of the Hölder inequality.

Remark 1.2 Note that this in particular means that the function f_s (replacing f_0 and f_1 in the classical situation (see (1))), mentioned in our preface can even be a member of a general measure space (X, μ) instead of the more simple cases mentioned there. And, moreover, that the continuous version of (1) has the form (1.3), i.e., also as an inequality between two geometric means but now in a much more general form.

It is also well-known that many of the classical inequalities hold also in a refined form, usually by adding some positive term to the smaller left-hand side. Indeed, we also prove that such refinements can be stated in this more general continuous/family frame we are working in.

The first time the present authors met the need to develop inequalities (mostly of Hölder type) in such continuous/family forms was in connection to that they were involved in the development of creating an interpolation theory between a family of/continuously many Banach spaces. Note that classical interpolation theory concerns usually interpolation between only two (or at most finitely many) Banach spaces. There are many similarities between these theories, e.g., the one pointed out above: to generalize inequalities for (interpolation between) two functions (Banach spaces) to cover a full continuous/family situation. These thoughts have influenced the text on some places and especially of Appendix, where we have given our readers a very short introduction how interpolation between families of Banach spaces can be understood for a broader audience.

The close relation between interpolation and inequalities is clearly illustrated by only considering the classical Riesz-Thorin interpolation theorem (see Theorem A.1 in Appendix). This close relation is also kept in more abstract situations, which we also explain shortly in our Appendix.

Another guiding idea for us is convexity. It is known that both convexity (almost equivalent to Jensen's inequality) and interpolation are very powerful tools to prove inequalities. And, more important, there are close connections between all of these three concepts (inequalities, interpolation, and convexity). And, most important, some of these relations are also partly kept in our continuous/family situation. Also, this fact has influenced our way of thinking and presenting in this book, see especially our Appendix.

In this Chap. 1, we state, prove, and discuss some continuous forms of the Hölder, Minkowski, Beckenbach-Dresher, Popoviciu, and Bellman inequalities. Moreover, a close (surprising) relation between the Hölder and Minkowski inequalities on these continuous/family forms is proved, namely they are in a sense equivalent. Also, a connection to the Gauss-Pòlya inequality is pointed out.

Chapter 2 is devoted to the corresponding generalizations in continuous/family direction of some refined inequalities. In particular, refined continuous forms of the inequalities by Hölder, Popoviciu, Minkowski, Bellman, Jensen, Jensen-Mercer and Hardy are stated and proved.

Chapter 3 is used to prove such improvements by using the concepts strong convexity and superquadraticity, two important notions connected to usual convexity. In particular, in this way some refined versions of continuous forms of the following inequalities are stated, proved, and applied: Jensen's inequality, Hermite-Hadamard's inequality, Minkowski's inequality, and Beckenbach-Dresher's inequality.

In Chap. 4 we carefully study the properties of some functionals associated to some of our inequalities in continuous/family forms. Typically, these functionals can be the "gaps" between the right-hand and left-hand sides of the inequalities. The properties of these functionals we prove are not only interesting in themselves but give also additional improvements (e.g., refinements) of the corresponding inequalities.

In Chap. 5 we focus on inequalities involving Banach lattice norms. We state and prove some new such Hölder-type, Minkowski-type, Popoviciu-type, Bellman-type, Beckenbach-Dresher-type, and Hardy-type inequalities, sometimes even combined with a full family situation.

Finally, in Appendix we briefly describe the close connection between power of inequalities, interpolation, and convexity in the classical situation. We also indicate that parts of these close connections are kept in our more general continuous/family situation. Moreover, we give a brief introduction to interpolation between families of Banach spaces which cannot be found in any of the traditional interpolation books, see for example, [17,18,22,59,122], etc. Finally, as an example that inequalities can hold also in this general frame, we finish the book by stating and proving a new Popoviciu-type inequality in the case of an interpolation family.

In this book we use the following notations: The measure space (X, Σ, μ) is often notated by (X, μ) or, only by X, when it is clear which positive measure is associated with X (see [118, p. 16]); the set of all measurable functions on the measure space (X, Σ, μ) will be denoted by $L_0(X)$ or by $L_0(X, \mu)$; the set of all integrable functions on the measure space (X, Σ, μ) will be denoted by $L_1(X)$ or by $L_1(X, \mu)$. Also, for brevity, we shall sometimes use the symbol $\int_X f\,\mathrm{d}\mu$ for the integral $\int_X f(x)\,\mathrm{d}\mu(x)$. For $p > 0$ by $L_p(X)$ or by $L_p(X, \mu)$ we denote a set of all functions $f : X \to \mathbb{R}$ for which $\int_X |f(x)|^p \mathrm{d}\mu(x)$ is finite. A weight function is a nonnegative, measurable function.

1.2 Continuous Forms of the Hölder and Minkowski Inequalities

1.2.1 On the Hölder Inequality

This section is devoted to the continuous form of the Hölder inequality and its consequences.

Theorem 1.1 *Let u and v be weight functions on the measure spaces (X, μ) and (Y, ν), respectively, such that $\int_X u(x)\,\mathrm{d}\mu(x) = 1$. Let f be a positive function on $X \times Y$ and measurable with respect to the measure $\mu \times \nu$. Then*

$$\int_Y \exp\left(\int_X \log f(x, y) u(x)\,\mathrm{d}\mu(x)\right) v(y)\,\mathrm{d}\nu(y)$$
$$\leq \exp\left(\int_X \log\left(\int_Y f(x, y) v(y)\,\mathrm{d}\nu(y)\right) u(x)\,\mathrm{d}\mu(x)\right). \quad (1.4)$$

Proof Let us recall the integral Jensen inequality, [85, p. 30]: If $g : X \to \mathbb{R}$ is an integrable function on the measure space (X, μ), u is a weight, and if φ is a convex function given on an interval I that includes the image of g, then $\frac{1}{\int_X u(x)\,\mathrm{d}\mu(x)} \int_X g(x) u(x)\,\mathrm{d}\mu(x) \in I$ when $0 < \int_X u(x)\,\mathrm{d}\mu(x) < \infty$ and

$$\varphi\left(\frac{1}{\int_X u(x)\,\mathrm{d}\mu(x)} \int_X g(x) u(x)\,\mathrm{d}\mu(x)\right)$$
$$\leq \frac{1}{\int_X u(x)\,\mathrm{d}\mu(x)} \int_X \varphi(g(x)) u(x)\,\mathrm{d}\mu(x). \quad (1.5)$$

Applying the Jensen inequality (1.5) for exponential function and using the assumptions, we get the following:

$$\exp\left(\int_X \log \frac{f(x, y)}{\int_Y f(x, y) v(y)\,\mathrm{d}\nu(y)} u(x)\,\mathrm{d}\mu(x)\right)$$
$$\leq \frac{1}{\int_X u(x)\,\mathrm{d}\mu(x)} \int_X \exp\left(\log \frac{f(x, y)}{\int_Y f(x, y) v(y)\,\mathrm{d}\nu(y)}\right) u(x)\,\mathrm{d}\mu(x)$$
$$= \int_X \frac{f(x, y)}{\int_Y f(x, y) v(y)\,\mathrm{d}\nu(y)} u(x)\,\mathrm{d}\mu(x).$$

Multiplying the above inequality with $v(y)$, integrating it over Y, and using the Fubini theorem, we get

$$\int_Y \exp\left(\int_X \log \frac{f(x,y)}{\int_Y f(x,y)v(y)\,dv(y)} u(x)\,d\mu(x)\right) v(y)\,dv(y)$$

$$\leq \int_Y \left(\int_X \frac{f(x,y)}{\int_Y f(x,y)v(y)\,dv(y)} u(x)\,d\mu(x)\right) v(y)\,dv(y)$$

$$= \int_X \left(\int_Y \frac{f(x,y)}{\int_Y f(x,y)v(y)\,dv(y)} v(y)\,dv(y)\right) u(x)\,d\mu(x)$$

$$= \int_X u(x)\,d\mu(x) = 1.$$

Since

$$\int_Y \exp\left(\int_X \log \frac{f(x,y)}{\int_Y f(x,y)v(y)\,dv(y)} u(x)\,d\mu(x)\right) v(y)\,dv(y)$$

$$= \frac{\int_Y \exp\left(\int_X \log f(x,y)u(x)\,d\mu(x)\right) v(y)\,dv(y)}{\exp\left(\int_X \log \left(\int_Y f(x,y)v(y)\,dv(y)\right) u(x)\,d\mu(x)\right)},$$

we observe that (1.4) holds and the proof is complete. □

Note that this proof is based on usage of convexity of exponential function, on application of the Jensen inequality for it and usage of the Fubini theorem. In [71], we can find a proof based on usage of the integral Minkowski inequality and the Fubini theorem. In the Sect. 1.2.3 we discuss the relationship between the Minkowski inequality and the Hölder inequality.

In the sequel we consider a useful consequence of Theorem 1.1.

Corollary 1.1 *Let* $w_1, \ldots, w_m \geq 0$ *be real numbers and let* $u \geq 0$, $p > 0$, $a_i > 0$ ($i = 1, 2, \ldots, m$) *be functions on X such that* $\int_X \frac{u(x)}{p(x)} d\mu(x) = 1$ *and* a_i^p *are measurable on X. Then*

$$\sum_{i=1}^m w_i \exp\left(\int_X \log a_i(x) u(x)\,d\mu(x)\right)$$

$$\leq \exp\left(\int_X \log\left(\sum_{i=1}^m w_i a_i^{p(x)}(x)\right) \frac{u(x)}{p(x)} d\mu(x)\right). \quad (1.6)$$

1.2 Continuous Forms of the Hölder and Minkowski Inequalities

Proof Without loss of generality we can assume that all w_i are positive. Put in Theorem 1.1:

$$Y = [0, m), \ Y_i = [i-1, i), \ i = 1, \ldots, m, \ f(x, y) = w_i a_i^{p(x)}(x) \text{ for } y \in Y_i,$$

$v(y) = 1$, $dv(y) = dy$, and $\frac{u(x)}{p(x)}$ instead of $u(x)$. Then the left-hand side of (1.4) is transformed into the following:

$$\int_Y \exp\left(\int_X \log f(x, y) \frac{u(x)}{p(x)} d\mu(x)\right) dy$$

$$= \sum_{i=1}^m \int_{Y_i} \exp\left(\int_X \log\left(w_i a_i^{p(x)}(x)\right) \frac{u(x)}{p(x)} d\mu(x)\right) dy$$

$$= \sum_{i=1}^m \exp\left(\int_X \log\left(w_i a_i^{p(x)}(x)\right) \frac{u(x)}{p(x)} d\mu(x)\right)$$

$$= \sum_{i=1}^m \exp\left(\int_X \log w_i \frac{u(x)}{p(x)} d\mu(x) + \int_X \log a_i(x) u(x) d\mu(x)\right)$$

$$= \sum_{i=1}^m w_i \exp\left(\int_X \log a_i(x) u(x) d\mu(x)\right). \tag{1.7}$$

The right-hand side of inequality (1.4) becomes

$$\exp\left(\int_X \log\left(\int_Y f(x, y) dy\right) \frac{u(x)}{p(x)} d\mu(x)\right)$$

$$= \exp\left(\int_X \log\left(\sum_{i=1}^m \int_{i-1}^i w_i a_i^{p(x)}(x) dy\right) \frac{u(x)}{p(x)} d\mu(x)\right)$$

$$= \exp\left(\int_X \log\left(\sum_{i=1}^m w_i a_i^{p(x)}(x)\right) \frac{u(x)}{p(x)} d\mu(x)\right). \tag{1.8}$$

From (1.7) and (1.8) we conclude inequality (1.6) and the proof is complete. □

Remark 1.3 Putting in Corollary 1.1: $m = 2$, $w_1 = w_2 = 1$, $p(x) = 1$ we get the following result, which can be described as superadditivity of geometric means:

$$\exp\left(\int_X \log a_1(x) u(x) \, d\mu(x)\right) + \exp\left(\int_X \log a_2(x) u(x) \, d\mu(x)\right)$$
$$\leq \exp\left(\int_X \log\bigl(a_1(x) + a_2(x)\bigr) u(x) \, d\mu(x)\right). \tag{1.9}$$

The above result can be found in [38, VI.11.35] and [97], but now we have shown how it follows from (1.4).

Example 1.2 Let $N = 2, 3, \ldots$, $X = [0,1]$, $d\mu(x) = dx$, $u \equiv 1$, $p_i > 1$ for $i = 1, 2, \ldots, N$, with $\sum_{i=1}^{N} \frac{1}{p_i} = 1$,

$$f(x, y) = \begin{cases} f_1^{p_1}(y), & 0 \leq x \leq \dfrac{1}{p_1} \\ f_2^{p_2}(y), & \dfrac{1}{p_1} < x \leq \dfrac{1}{p_1} + \dfrac{1}{p_2} \\ \vdots \\ f_N^{p_N}(y), & \sum_{i=1}^{N-1} \dfrac{1}{p_i} < x \leq 1. \end{cases}$$

Then (1.4) reads

$$\int_Y \prod_{i=1}^{N} f_i(y) v(y) \, d\nu(y) \leq \prod_{i=1}^{N} \left(\int_Y f_i^{p_i}(y) v(y) \, d\nu(y)\right)^{1/p_i},$$

i.e., it is a standard form of the Hölder inequality involving N functions.

1.2.2 On the Minkowski Inequality

Another well-known classical inequality is the Minkowski inequality. In the literature we can find several Minkowski-type inequalities such as the discrete Minkowski inequality for two or several real sequences, the Minkowski inequality for integrals involving two or several functions, the Minkowski inequality for isotonic linear functional, etc. Here we will focus on the continuous version of the Minkowski inequality (often called the integral form of the Minkowski inequality), see for example [73, p. 41].

Theorem 1.2 Let $f(x, y)$ be nonnegative and measurable on $(X \times Y, \mu \times \nu)$ and let $u(x)$ and $v(y)$ be weight functions.

1.2 Continuous Forms of the Hölder and Minkowski Inequalities

(a) If $p \geq 1$, then

$$\left(\int_Y \left(\int_X f(x,y)u(x)\,d\mu(x)\right)^p v(y)\,dv(y)\right)^{\frac{1}{p}}$$
$$\leq \int_X \left(\int_Y f^p(x,y)v(y)\,dv(y)\right)^{\frac{1}{p}} u(x)\,d\mu(x). \quad (1.10)$$

(b) If $0 < p < 1$ and:

(i) $\int_X \left(\int_Y f(x,y)v(y)\,dv(y)\right)^p u(x)\,d\mu(x) > 0$ (μ-a.e.) and

$\int_Y f(x,y)v(y)\,dv(y) > 0$ (v-a.e.), then the reverse inequality in (1.10) holds.

If $p < 0$, the above-mentioned assumptions (i) and the additional one:

(ii) $\int_X f^p(x,y)u(x)\,d\mu(x) > 0$ (v-a.e.)

hold, then the reverse inequality in (1.10) holds.

Here we present two proofs of Theorem 1.2. The first one is based on *quasilinearization method* and on application of the Fubini theorem, while in the second one, we use the Hölder inequality for integrals.

Proof *The first proof of (a)*. Let us mention (again) the Hölder inequality for integrals involving two positive functions f_1 and f_2 (see (1.2)):

$$\int_Y f_1(y)f_2(y)v(y)\,dv(y) \leq \left(\int_Y f_1^p(y)v(y)\,dv(y)\right)^{\frac{1}{p}} \left(\int_Y f_2^q(y)v(y)\,dv(y)\right)^{\frac{1}{q}}, \quad (1.11)$$

where $p > 1$, $\frac{1}{p} + \frac{1}{q} = 1$. It is obvious that equality holds in it if f_1 and f_2 are functions such that $f_2^q = f_1^p$.

This fact gives the following representation (quasilinearization) formula:

$$\|f_1\|_{p,v} = \sup_{f_2} \int_Y f_2(y)f_1(y)v(y)\,dv(y), \quad (1.12)$$

where

$$\|f_1\|_{p,v} = \left(\int_Y f_1^p(y)v(y)\,dv(y)\right)^{\frac{1}{p}}, \quad p > 1,$$

and supremum is taken over all $f_2 \in L_q(Y)$, $\frac{1}{p} + \frac{1}{q} = 1$ so that $\|f_2\|_{q,v} = 1$.

By using the representation formula (1.12) and the Fubini theorem, we find that

$$\left(\int_Y \left(\int_X f(x,y)u(x)\,d\mu(x)\right)^p v(y)\,dv(y)\right)^{\frac{1}{p}} = \left\|\int_X f(x,y)u(x)\,d\mu(x)\right\|_{p,v}$$

$$= \sup_{\|g\|_{q,v}=1} \int_Y g(y) \left(\int_X f(x,y)u(x)\,d\mu(x)\right) v(y)\,dv(y)$$

$$= \sup_{\|g\|_{q,v}=1} \int_X u(x) \left(\int_Y g(y)f(x,y)v(y)\,dv(y)\right) d\mu(x)$$

$$\leq \int_X \sup_{\|g\|_{q,v}=1} \left(\int_Y g(y)f(x,y)v(y)\,dv(y)\right) u(x)\,d\mu(x)$$

$$= \int_X \left(\int_Y f^p(x,y)v(y)\,dv(y)\right)^{\frac{1}{p}} u(x)\,d\mu(x).$$

The second proof of (a). If $p = 1$, then the statement is obvious—it is just the Fubini theorem. Let us suppose $p > 1$ and define q as $\frac{1}{q} = \frac{p-1}{p}$. Let us denote

$$F(y) = \int_X f(x,y)u(x)d\mu(x).$$

Without loss of generality we assume that the left-hand side of (1.10) is positive. Simple transformations, usage of the Fubini theorem, and the integral Hölder inequality for two functions give the following:

$$\int_Y F^p(y)v(y)dv(y) = \int_Y F(y)F^{p-1}(y)v(y)dv(y)$$

$$= \int_Y \left(\int_X f(x,y)u(x)d\mu(x)\right) F^{p-1}(y)v(y)dv(y)$$

$$= \int_X \left(\int_Y f(x,y)F^{p-1}(y)v(y)dv(y)\right) u(x)d\mu(x)$$

$$\leq \int_X \left(\int_Y f^p(x,y)v(y)dv(y)\right)^{\frac{1}{p}} \left(\int_Y F^p(y)v(y)dv(y)\right)^{\frac{1}{q}} u(x)d\mu(x)$$

$$= \left(\int_Y F^p(y)v(y)dv(y)\right)^{\frac{1}{q}} \cdot \int_X \left(\int_Y f^p(x,y)v(y)dv(y)\right)^{\frac{1}{p}} u(x)d\mu(x).$$

Dividing the above inequality with $\left(\int_Y F^p(y)v(y)dv(y)\right)^{\frac{1}{q}}$, we get (1.10).

Proof of (b). In this case the Hölder inequality (1.11) holds in the reversed direction but still has equality if $f_1^p = f_2^q$. So in this case the quasilinearization formula (1.12) holds with "sup" replaced by "inf." Hence, the proof follows in a similar way as of (a) so we omit details. The proof is complete. □

Example 1.3 Let $X_1, \ldots, X_n$ be a partition of X, $\int_{X_i} u(x)\,d\mu(x) = \alpha_i$, and $f(x, y) = \frac{f_i(y)}{\alpha_i}$, for $x \in X_i$, $(i = 1, 2, \ldots, n)$.

If $p \geq 1$, then (1.10) becomes

$$\left(\int_Y \left(\sum_{i=1}^n f_i(y)\right)^p v(y)\,d\nu(y)\right)^{\frac{1}{p}} \leq \sum_{i=1}^n \left(\int_Y f_i^p(y) v(y)\,d\nu(y)\right)^{\frac{1}{p}}, \quad (1.13)$$

i.e., the usual Minkowski inequality for integrals with n nonnegative functions involved.

If $0 < p < 1$ or $p < 0$, then, with the obvious restrictions on the integrals, (1.13) holds in the reversed direction.

Similarly, as a consequence of Theorem 1.2 we obtain the classical discrete Minkowski inequality. Namely, the following inequality holds for nonnegative real m-tuples $(w_1, \ldots, w_m)$, $(a_{i1}, \ldots, a_{im})$ $(i = 1, \ldots n)$ and $p \geq 1$:

$$\left(\sum_{j=1}^m w_j (a_{1j} + \ldots + a_{nj})^p\right)^{1/p}$$
$$\leq \left(\sum_{j=1}^m w_j a_{1j}^p\right)^{1/p} + \ldots + \left(\sum_{j=1}^m w_j a_{nj}^p\right)^{1/p}. \quad (1.14)$$

If $0 < p < 1$ or if $p < 0$ with $\sum_{j=1}^m w_j a_{1j}^p > 0$, $\ldots$, $\sum_{j=1}^m w_j a_{nj}^p > 0$, then the reverse inequality in (1.14) holds.

1.2.3 Relation Between the Hölder and Minkowski Inequalities

Let us point out an important relation between inequalities (1.4) and (1.10) via the so-called power means.

Let $\int_X u(x)\,d\mu(x) = 1$. Denote (the power means)

$$M^{[r]}(f, \mu, u) = \left(\int_X f^r(x, y) u(x)\,d\mu(x)\right)^{\frac{1}{r}}, \quad r \neq 0$$

and

$$M^{[0]}(f, \mu, u) = \exp\left(\int_X \log f(x, y) u(x) \, d\mu(x)\right).$$

It is well-known that $M^{[r]}(f, \mu, u)$ is a nondecreasing function on r (see [85, p. 33]) and, moreover,

$$\lim_{r \to 0} M^{[r]}(f, \mu, u) = M^{[0]}(f, \mu, u).$$

Next we prove that (1.10) implies (1.4). In fact, let us put in (1.10) $f^{\frac{1}{p}}(x, y)$ instead of $f(x, y)$ and take both sides of the inequality in power p. We get

$$\int_Y M^{[\frac{1}{p}]}(f, \mu, u) v(y) \, dv(y) \leq M^{[\frac{1}{p}]}(h, \mu, u),$$

where $h(x) = \int_Y f(x, y) v(y) \, dv(y)$.

Let $p \to \infty$. Then, since $M^{[\frac{1}{p}]}$ monotonously decreases in p, the left-hand side goes to $\int_Y M^{[0]}(f, \mu, u) v(y) \, dv(y)$ by the monotone convergence theorem. This gives us the left-hand side of (1.4). Since

$$M^{[\frac{1}{p}]}(h, \mu, u) \to M^{[0]}(h, \mu, u)$$
$$= \exp\left(\int_X \log\left(\int_Y f(x, y) v(y) \, dv(y)\right) u(x) \, d\mu(x)\right),$$

we get inequality (1.4).

However, our proof in the previous section shows that, in a sense, the remarkable fact that the continuous versions of the inequalities are in fact equivalent (see [97]).

Theorem 1.3 *Let $f(x, y)$ be positive and measurable on $(X \times Y, \mu \times \nu)$. Assume that $p \geq 1$ and that $u(x)$ and $v(y)$ are weight functions. Then the following statements are equivalent:*

(i) *The continuous Hölder inequality (1.4) holds for all X, $u(x)$ and $\mu(x)$ such that $\int_X u(x) \, d\mu(x) = 1$.*
(ii) *The continuous Minkowski inequality (1.10) holds for all X, $u(x)$ and $\mu(x)$ such that $\int_X u(x) \, d\mu(x) < \infty$.*

Proof Assume that (ii) holds. Then, by using our limit argument (via power means), we have proved just before this theorem that (i) holds.

Assume that (i) holds, and use this in the special case to get the Hölder integral inequality for two functions, so that the quasilinearization formula (1.12) holds. Then, as proved in Sect. 1.2.2, (ii) holds. The proof is complete. □

1.2.4 On the Beckenbach-Dresher Inequality

An inequality which is closely related to the Hölder and Minkowski inequality is the so-called Beckenbach-Dresher inequality. The first version of that inequality was published by E. F. Beckenbach in [13]. He stated that for positive real numbers $x_i, y_i, i = 1, \ldots, n$ and for $1 \leq p \leq 2$ the following inequality

$$\frac{\sum_{i=1}^{n}(x_i + y_i)^p}{\sum_{i=1}^{n}(x_i + y_i)^{p-1}} \leq \frac{\sum_{i=1}^{n} x_i^p}{\sum_{i=1}^{n} x_i^{p-1}} + \frac{\sum_{i=1}^{n} y_i^p}{\sum_{i=1}^{n} y_i^{p-1}} \tag{1.15}$$

is valid. If $0 \leq p \leq 1$, then inequality (1.15) is reversed.

Few years later M. Dresher (see [37]) investigated moment spaces and stated that an integral analogue of the previous result holds. In fact, he proved that if $p \geq 1 \geq q \geq 0$, and $f, g \geq 0$, then

$$\left(\frac{\int (f+g)^p d\varphi}{\int (f+g)^q d\varphi} \right)^{\frac{1}{p-q}} \leq \left(\frac{\int f^p d\varphi}{\int f^q d\varphi} \right)^{\frac{1}{p-q}} + \left(\frac{\int g^p d\varphi}{\int g^q d\varphi} \right)^{\frac{1}{p-q}}. \tag{1.16}$$

From a large number of generalizations of (1.16) (see [19, 45, 90, 107, 111, 112, 123] and references therein), we will highlight a version given by J. Peetre and L.-E. Persson in [111], which involves three parameters u, p and q.

Theorem 1.4 *Let $A, B : L \to \mathbb{R}$ be two positive linear functionals and $f_i, u_i : E \to [0, \infty)$, $(i = 1, \ldots, n)$, be functions such that $f_i^p, f_i^q, (\sum_{i=1}^{n} f_i)^p, (\sum_{i=1}^{n} f_i)^q, \in L$, $(i = 1, \ldots, n)$.*

If $u \geq 1$ and $q \leq 1 \leq p$ $(q \neq 0)$, then

$$\frac{A^{\frac{u}{p}}\left(\left(\sum_{i=1}^{n} f_i\right)^p\right)}{B^{\frac{u-1}{q}}\left(\left(\sum_{i=1}^{n} f_i\right)^q\right)} \leq \sum_{i=1}^{n} \frac{A^{\frac{u}{p}}(f_i^p)}{B^{\frac{u-1}{q}}(f_i^q)}. \tag{1.17}$$

If $0 < u \leq 1$, $p \leq 1$, and $q \leq 1$, $p, q \neq 0$, then inequality (1.17) is reversed.

The proof of the above result is based on the following property of convex increasing function $F : \mathbb{R}_+^n \to \mathbb{R}_+$, superadditive function $g : D \to \mathbb{R}_+$, and subadditive function $f : D \to \mathbb{R}_+^n$ (see [81, 111]):

$$g(x+y)F\left(\frac{f(x+y)}{g(x+y)}\right) \leq g(x)F\left(\frac{f(x)}{g(x)}\right) + g(y)F\left(\frac{f(y)}{g(y)}\right).$$

Also, we mention here a result from [45] which is in fact a continuous form of the Beckenbach-Dresher inequality.

Theorem 1.5 *Let (X, μ), (Y, ν), and (Y, λ) be measure spaces. Let f, g be nonnegative functions on $X \times Y$ such that f is integrable with respect to the measure $\mu \times \nu$ and g is integrable with respect to $\mu \times \lambda$.*

(a) If:
  (i) $u \geq 1$ and $q \leq 1 \leq p$ ($q \neq 0$), or
  (ii) $u < 0$ and $p \leq 1 \leq q$ ($p \neq 0$), and all terms exist, then

$$\frac{\left(\int_Y \left(\int_X f(x,y) d\mu(x)\right)^p d\nu(y)\right)^{\frac{u}{p}}}{\left(\int_Y \left(\int_X g(x,y) d\mu(x)\right)^q d\lambda(y)\right)^{\frac{u-1}{q}}} \leq \int_X \frac{\left(\int_Y f^p(x,y) d\nu(y)\right)^{\frac{u}{p}}}{\left(\int_Y g^q(x,y) d\lambda(y)\right)^{\frac{u-1}{q}}} d\mu(x). \tag{1.18}$$

If (iii) $0 < u \leq 1$, $p \leq 1$, and $q \leq 1$, $p, q \neq 0$, then inequality (1.18) is reversed.
(b) If $u \geq 1$ and $p \geq 1$, then

$$\left(\int_Y \left(\int_X f(x,y) d\mu(x)\right)^p d\nu(y)\right)^{\frac{u}{p}} \times$$

$$\times \exp\left(\frac{1-u}{\int_Y d\lambda} \int_Y \log\left(\int_X g(x,y) d\mu(x)\right) d\lambda(y)\right)$$

$$\leq \int_X \left(\int_Y f^p(x,y) d\nu(y)\right)^{\frac{u}{p}} \exp\left(\frac{1-u}{\int_Y d\lambda} \int_Y \log g(x,y) d\lambda(y)\right) d\mu(x).$$

Remark 1.4 Using a discrete measure μ in (1.18), we can formulate the following Beckenbach-Dresher inequality (of Persson-Peetre type) for integrals (see [45, 123]): If u, p, q satisfy the assumptions of Theorem 1.5(i) or (ii), then

$$\frac{\left(\int_Y \left(\sum_{i=1}^n f_i(y)\right)^p d\nu(y)\right)^{\frac{u}{p}}}{\left(\int_Y \left(\sum_{i=1}^n g_i(y)\right)^q d\lambda(y)\right)^{\frac{u-1}{q}}} \leq \sum_{i=1}^n \frac{\left(\int_Y f_i^p(y) d\nu(y)\right)^{\frac{u}{p}}}{\left(\int_Y g_i^q(y) d\lambda(y)\right)^{\frac{u-1}{q}}}. \quad (1.19)$$

Moreover, (1.19) holds in the reversed direction with the corresponding restrictions of the parameters.

1.3 Continuous Forms of the Popoviciu and Bellman Inequalities

1.3.1 On the Popoviciu Inequality

The classical discrete Hölder inequality, i.e., the Hölder inequality for n-tuples, involves nonnegative weights. In the middle of the twentieth century, T. Popoviciu (see [116]) studied Hölder-type inequalities for other types of weights and proved results that we now call Popoviciu's inequalities or the reversed Hölder inequalities or the Hölder-Lorentz inequalities (see [24, p.199]). With regard to the objects that are in the inequality, we distinguish between the discrete Popoviciu inequality, the Popoviciu inequality for integrals, and the Popoviciu inequality for isotonic linear functionals. In this section we present results about the continuous form of this inequality which were published in [97] (see also [7]).

For the reader convenience we will state and prove such inequalities for n-tuples and for integrals. More results related to this problem can be found in the very rich literature, see for example [24, 75, 110, 126] and references therein.

Theorem 1.6 *Let p, q be positive real numbers such that $\frac{1}{p} + \frac{1}{q} = 1$.*

(i) Let $v_0, c_1, c_2 > 0$, $v_i, f_i, g_i \geq 0$, $i = 1, 2, \ldots, n$ be real numbers such that $v_0 c_1^p - \sum_{i=1}^n v_i f_i^p > 0$, $v_0 c_2^q - \sum_{i=1}^n v_i g_i^q > 0$. Then

$$v_0 c_1 c_2 - \sum_{i=1}^n v_i f_i g_i \geq \left(v_0 c_1^p - \sum_{i=1}^n v_i f_i^p\right)^{\frac{1}{p}} \left(v_0 c_2^q - \sum_{i=1}^n v_i g_i^q\right)^{\frac{1}{q}}. \quad (1.20)$$

(ii) Let f and g be nonnegative measurable functions on the measure space (Y, ν). Then the following inequality

$$v_0 c_1 c_2 - \int_Y v(y) f(y) g(y) \, d\nu(y)$$
$$\geq \left(v_0 c_1^p - \int_Y v(y) f^p(y) \, d\nu(y) \right)^{\frac{1}{p}} \left(v_0 c_2^q - \int_Y v(y) g^q(y) \, d\nu(y) \right)^{\frac{1}{q}} \quad (1.21)$$

holds, where $v(y)$ is a weight, $v_0 c_1^p - \int_Y v(y) f^p(y) \, d\nu(y) \geq 0$, and $v_0 c_2^q - \int_Y v(y) g^q(y) \, d\nu(y) \geq 0$.

Proof Let us first prove (ii) part. Put in the discrete Hölder inequality with nonnegative weights w_1 and w_2:

$$(w_1 a_1^p + w_2 a_2^p)^{1/p} (w_1 b_1^q + w_2 b_2^q)^{1/q} \geq w_1 a_1 b_1 + w_2 a_2 b_2, \quad (1.22)$$

where $a_1, a_2, b_1, b_2 \geq 0$, the following expressions:

$$w_1 a_1^p = v_0 c_1^p - \int_Y v(y) f^p(y) \, d\nu(y), \quad w_1 b_1^q = v_0 c_2^q - \int_Y v(y) g^q(y) \, d\nu(y),$$

$$w_2 a_2^p = \int_Y v(y) f^p(y) \, d\nu(y) \quad \text{and} \quad w_2 b_2^q = \int_Y v(y) g^q(y) \, d\nu(y).$$

Then the left-hand side of (1.22) collapses to $v_0 c_1 c_2$, and the right-hand side of (1.22) becomes equal to

$$\left(v_0 c_1^p - \int_Y v(y) f^p(y) \, d\nu(y) \right)^{1/p} \left(v_0 c_2^q - \int_Y v(y) g^q(y) \, d\nu(y) \right)^{1/q}$$
$$+ \left(\int_Y v(y) f^p(y) \, d\nu(y) \right)^{1/p} \left(\int_Y v(y) g^q(y) \, d\nu(y) \right)^{1/q}$$
$$\geq \left(v_0 c_1^p - \int_Y v(y) f^p(y) \, d\nu(y) \right)^{1/p} \left(v_0 c_2^q - \int_Y v(y) g^q(y) \, d\nu(y) \right)^{1/q}$$
$$+ \int_Y v(y) f(y) g(y) \, d\nu(y),$$

where in the last inequality we have used the Hölder inequality for integrals. After a simple transformation we get (1.21). Furthermore, applying (1.21) on a discrete measure, we obtain (1.20). So the proof is complete. □

1.3 Continuous Forms of the Popoviciu and Bellman Inequalities

Next we present a generalization of (1.21), namely the following continuous form of it (see [97]):

Theorem 1.7 *Let $u(x)$ and $v(y)$ be weight functions on the measure spaces (X, μ) and (Y, ν), respectively, such that $\int_X u(x) \, d\mu(x) = 1$, let $f(x, y)$ be a positive measurable function on $X \times Y$, $v_0 \in (0, \infty)$, and assume that $f_0(x)$ is a function on X such that $v_0 f_0(x) > \int_Y f(x, y) v(y) \, d\nu(y)$, for all $x \in X$. Then the following continuous form of the Popoviciu inequality holds:*

$$\exp\left(\int_X \log(v_0 f_0(x)) u(x) \, d\mu(x)\right)$$
$$- \int_Y \exp\left(\int_X \log f(x, y) u(x) \, d\mu(x)\right) v(y) \, d\nu(y)$$
$$\geq \exp\left[\int_X \log\left(v_0 f_0(x) - \int_Y f(x, y) v(y) \, d\nu(y)\right) u(x) \, d\mu(x)\right]. \quad (1.23)$$

Proof Let us remind that in Sect. 1.2.1 we proved the following inequality (see (1.9))

$$\exp\left(\int_X \log a_1(x) u(x) \, d\mu(x)\right) + \exp\left(\int_X \log a_2(x) u(x) \, d\mu(x)\right)$$
$$\leq \exp\left(\int_X \log\left(a_1(x) + a_2(x)\right) u(x) \, d\mu(x)\right), \quad (1.24)$$

where u, a_1, a_2 are positive functions on X with $\int_X u(x) \, d\mu(x) = 1$.

Applying inequality (1.24) for

$$a_1(x) = v_0 f_0(x) - \int_Y f(x, y) v(y) \, d\nu(y) \text{ and } a_2(x) = \int_Y f(x, y) v(y) \, d\nu(y),$$

we get

$$\exp\left(\int_X \log(v_0 f_0(x)) u(x) \, d\mu(x)\right)$$
$$\geq \exp\left(\int_X \log\left(\int_Y f(x, y) v(y) \, d\nu(y)\right) u(x) \, d\mu(x)\right)$$
$$+ \exp\left[\int_X \log\left(v_0 f_0(x) - \int_Y f(x, y) v(y) \, d\nu(y)\right) u(x) \, d\mu(x)\right]$$
$$\geq \int_Y \exp\left(\int_X \log f(x, y) u(x) \, d\mu(x)\right) v(y) \, d\nu(y)$$

$$+ \exp\left[\int_X \log\left(v_0 f_0(x) - \int_Y f(x, y)v(y)\,dv(y)\right) u(x)\,d\mu(x)\right],$$

so that (1.23) holds. In the last inequality we have used Theorem 1.1. The proof is complete. □

Example 1.4 Let $u(x) = 1$, $X = X_1 \cup X_2$, $X_1 \cap X_2 = \emptyset$ with $\int_{X_1} d\mu(x) = \frac{1}{p}$, $\int_{X_2} d\mu(x) = \frac{1}{q}$, where $\frac{1}{p} + \frac{1}{q} = 1$,

$$f_0(x) = \begin{cases} c_1^p, & x \in X_1 \\ c_2^q, & x \in X_2 \end{cases} \quad \text{and} \quad f(x, y) = \begin{cases} f^p(y), & x \in X_1 \\ g^q(y), & x \in X_2. \end{cases}$$

Then we rediscover the Popoviciu inequality (1.21) in the finite form with $v_0 = 1$.

The following result is also a consequence of the continuous form of the Popoviciu inequality, but it is also of independent interest in our future studying of continuous forms (see [92]).

Corollary 1.2 Let $w_1 > 0$, $w_2, \ldots, w_m \geq 0$ be reals, and $p, a_i, i = 1, 2, \ldots, m$, be positive functions on X such that $\int_X \frac{d\mu(x)}{p(x)} = 1$ and a_i^p are measurable on X. Then

$$w_1 \exp\left(\int_X \log a_1(x)\,d\mu(x)\right) - \sum_{i=2}^m w_i \exp\left(\int_X \log a_i(x)\,d\mu(x)\right)$$

$$\geq \exp\left\{\int_X \log\left[w_1(a_1(x))^{p(x)} - \sum_{i=2}^m w_i(a_i(x))^{p(x)}\right] \frac{d\mu(x)}{p(x)}\right\}, \quad (1.25)$$

provided that all integrals exist.

Proof Without loss of generality we can assume that all w_i are positive. Put in Theorem 1.7 related to the continuous Popoviciu inequality:

$$f_0(x) = w_1(a_1(x))^{p(x)}, \quad v_0 = 1, \quad u(x) = \frac{1}{p(x)}, \quad v(y) = 1, \quad dv(y) = dy,$$

$$Y = [1, m), \quad Y_i = [i-1, i), \quad i = 2, \ldots, m, \quad f(x, y) = w_i(a_i(x))^{p(x)} \text{ for } y \in Y_i.$$

1.3 Continuous Forms of the Popoviciu and Bellman Inequalities

Then

$$\exp\left\{\int_X \log\left[w_1(a_1(x))^{p(x)} - \sum_{i=2}^m w_i a_i(x)^{p(x)}\right] \frac{d\mu(x)}{p(x)}\right\}$$

$$= \exp\left[\int_X \log\left(f_0(x) - \int_Y f(x,y)v(y)\,dv(y)\right) u(x)\,d\mu(x)\right]$$

$$\leq \exp\left(\int_X \log f_0(x) u(x)\,d\mu(x)\right)$$

$$- \int_Y \exp\left(\int_X \log f(x,y) u(x)\,d\mu(x)\right) v(y)\,dv(y)$$

$$= w_1 \exp\left(\int_X \log a_1(x)\,d\mu(x)\right) - \sum_{i=2}^m w_i \exp\left(\int_X \log a_i(x)\,d\mu(x)\right),$$

which means that (1.25) holds and the proof is complete. □

Remark 1.5 By analyzing our proof of Theorem 1.7 we see that we can obtain a more general (a stronger) inequality than (1.23) by just not using the final estimate. We consider the functionals

$$P_1(v_0, v) = \exp\left(\int_X \log(v_0 f_0(x)) u(x)\,d\mu(x)\right)$$

$$- \int_Y \exp\left(\int_X \log f(x,y) u(x)\,d\mu(x)\right) v(y)\,dv(y)$$

$$- \exp\left[\int_X \log\left(v_0 f_0(x) - \int_Y f(x,y)v(y)\,dv(y)\right) u(x)\,d\mu(x)\right]$$

and

$$P_2(v_0, v) = \exp\left(\int_X \log(v_0 f_0(x)) u(x)\,d\mu(x)\right)$$

$$- \exp\left(\int_X \log\left(\int_Y f(x,y)v(y)\,dv(y)\right) u(x)\,d\mu(x)\right)$$

$$- \exp\left[\int_X \log\left(v_0 f_0(x) - \int_Y f(x,y)v(y)\,dv(y)\right) u(x)\,d\mu(x)\right].$$

In fact the proof of Theorem 1.7 shows that (1.23) can be replaced by the following (refined) version

$$P_1(v_0, v) \geq P_2(v_0, v) \geq 0. \qquad (1.26)$$

In particular, by applying (1.26) for the special case pointed out in Remark 1.4, we obtain the following refinement of the integral Popoviciu inequality (1.21), with $v_0 = 1$:

$$c_1 c_2 - \int_Y v(y) f(y) g(y) \, dv(y)$$

$$\geq c_1 c_2 - \left(\int_Y v(y) f^p(y) \, dv(y) \right)^{\frac{1}{p}} \left(\int_Y v(y) g^q(y) \, dv(y) \right)^{\frac{1}{q}}$$

$$\geq \left(c_1^p - \int_Y v(y) f^p(y) \, dv(y) \right)^{\frac{1}{p}} \left(c_2^q - \int_Y v(y) g^q(y) \, dv(y) \right)^{\frac{1}{q}}.$$

1.3.2 On the Bellman Inequality

The Bellman inequality is related to the Minkowski inequality in a similar way such as the Popoviciu inequality is connected to the Hölder inequality. It can be considered as a reverse version of the Minkowski inequality, and sometimes it is called the Minkowski-Lorentz inequality (see [24, p. 199]). Since its discovery by R. Bellman in [16], this inequality has undergone many generalizations (see for example [13, 110], and references therein). For the reader's convenience, here we state the Bellman inequality in its discrete and integral forms.

Theorem 1.8

(i) Let $v_0, c_1, c_2 > 0$, $v_i, f_i, g_i \geq 0$, $i = 1, 2, \ldots, n$ be real numbers such that $v_0 c_1^p - \sum_{i=1}^{n} v_i f_i^p \geq 0$, $v_0 c_2^p - \sum_{i=1}^{n} v_i g_i^p \geq 0$. If $p \geq 1$, then

$$\left(v_0 (c_1 + c_2)^p - \sum_{i=1}^{n} v_i (f_i + g_i)^p \right)^{\frac{1}{p}}$$

$$\geq \left(v_0 c_1^p - \sum_{i=1}^{n} v_i f_i^p \right)^{\frac{1}{p}} + \left(v_0 c_2^p - \sum_{i=1}^{n} v_i g_i^p \right)^{\frac{1}{p}}. \qquad (1.27)$$

1.3 Continuous Forms of the Popoviciu and Bellman Inequalities

(ii) Let $v_0, c_1, c_2 > 0$, f, g and v be nonnegative measurable functions on the measure space (Y, v) such that $v_0 c_1^p - \int_Y v(y) f^p(y) \, dv(y) \geq 0$ and $v_0 c_2^p - \int_Y v(y) g^p(y) \, dv(y) \geq 0$. If $p \geq 1$, then

$$\left(v_0(c_1+c_2)^p - \int_Y v(y)(f(y)+g(y))^p \, dv(y) \right)^{\frac{1}{p}}$$
$$\geq \left(v_0 c_1^p - \int_Y v(y) f^p(y) dv(y) \right)^{\frac{1}{p}} + \left(v_0 c_2^p - \int_Y v(y) g^p(y) dv(y) \right)^{\frac{1}{p}}. \quad (1.28)$$

Proof Let us first prove (ii) part. Put in the discrete Minkowski inequality for numbers $a_1, a_2, b_1, b_2 \geq 0$ and with nonnegative weights w_1, w_2

$$\left[(w_1 a_1^p + w_2 a_2^p)^{1/p} + (w_1 b_1^p + w_2 b_2^p)^{1/p} \right]^p \geq w_1(a_1+b_1)^p + w_2(a_2+b_2)^p$$

the following:

$$w_1 a_1^p = v_0 c_1^p - \int_Y v(y) f^p(y) \, dv(y), \quad w_2 a_2^p = \int_Y v(y) f^p(y) \, dv(y),$$

$$w_1 b_1^p = v_0 c_2^p - \int_Y v(y) g^p(y) \, dv(y), \text{ and } w_2 b_2^p = \int_Y v(y) g^p(y) \, dv(y).$$

Then we get

$$v_0(c_1+c_2)^p$$
$$\geq \left[\left(v_0 c_1^p - \int_Y v(y) f^p(y) \, dv(y) \right)^{\frac{1}{p}} + \left(v_0 c_2^p - \int_Y v(y) g^p(y) \, dv(y) \right)^{\frac{1}{p}} \right]^p$$
$$+ \left[\left(\int_Y v(y) f^p(y) \, dv(y) \right)^{\frac{1}{p}} + \left(\int_Y v(y) g^p(y) \, dv(y) \right)^{\frac{1}{p}} \right]^p$$
$$\geq \left[\left(v_0 c_1^p - \int_Y v(y) f^p(y) \, dv(y) \right)^{\frac{1}{p}} + \left(v_0 c_2^p - \int_Y v(y) g^p(y) \, dv(y) \right)^{\frac{1}{p}} \right]^p$$
$$+ \int_Y v(y)(f(y)+g(y))^p \, dv(y).$$

Taking the pth root from the above inequality, we obtain inequality (1.28). Moreover, applying (1.28) with a discrete measure, we can conclude that also (1.27) holds, so the proof is complete. □

In this section we present results about the continuous forms of these inequalities which were published in [97].

Theorem 1.9 *Let $u(x)$ and $v(y)$ be weight functions on the measure spaces (X, μ) and (Y, ν), respectively, $f(x, y)$ be a nonnegative measurable function on $X \times Y$, $v_0 \in (0, \infty)$, and assume that $f_0(x)$ is a nonnegative function on X. Then, for $p \geq 1$,*

$$\left(\int_X \left[v_0 f_0^p(x) - \int_Y f^p(x,y) v(y) \, d\nu(y) \right]^{\frac{1}{p}} u(x) \, d\mu(x) \right)^p$$
$$\leq v_0 \left[\int_X f_0(x) u(x) \, d\mu(x) \right]^p - \int_Y \left[\int_X f(x,y) u(x) \, d\mu(x) \right]^p v(y) \, d\nu(y), \quad (1.29)$$

whenever $v_0 f_0^p(x) \geq \int_Y f^p(x,y) v(y) \, d\nu(y)$, for all $x \in X$.

Proof First we apply Theorem 1.2 for the special case when $Y = Y_1 \bigcup Y_2$, $Y_1 \cap Y_2 = \emptyset$, $f(x,y) = a(x)$ on Y_1, $f(x,y) = b(x)$ on Y_2 and $\int_{Y_1} v(y) d\nu(y) = \int_{Y_2} v(y) d\nu(y) = 1$ and get that

$$\left(\int_X a(x) u(x) \, d\mu(x) \right)^p + \left(\int_X b(x) u(x) \, d\mu(x) \right)^p$$
$$\leq \left[\int_X \left(a^p(x) + b^p(x) \right)^{\frac{1}{p}} u(x) \, d\mu(x) \right]^p.$$

We choose

$$a(x) = \left[v_0 f_0^p(x) - \int_Y f^p(x,y) v(y) \, d\nu(y) \right]^{\frac{1}{p}},$$
$$b(x) = \left[\int_Y f^p(x,y) v(y) \, d\nu(y) \right]^{\frac{1}{p}}$$

and obtain that

$$\left(\int_X \left[v_0 f_0^p(x) - \int_Y f^p(x,y) v(y) \, d\nu(y) \right]^{\frac{1}{p}} u(x) \, d\mu(x) \right)^p$$
$$+ \left(\int_X \left[\int_Y f^p(x,y) v(y) d\nu(y) \right]^{\frac{1}{p}} u(x) d\mu(x) \right)^p$$
$$\leq v_0 \left[\int_X f_0(x) u(x) d\mu(x) \right]^p := I_1.$$

1.3 Continuous Forms of the Popoviciu and Bellman Inequalities

Next, by using Theorem 1.2 to the second term in the above inequality, we find that

$$I_1 \geq \int_Y \left(\int_X f(x,y) u(x) \, d\mu(x) \right)^p v(y) \, dv(y)$$
$$+ \left(\int_X \left[v_0 f_0^p(x) - \int_Y f^p(x,y) v(y) \, dv(y) \right]^{\frac{1}{p}} u(x) \, d\mu(x) \right)^p. \quad (1.30)$$

By using (1.30), in view of the definition of I_1, we obtain (1.29). The proof is complete. □

Example 1.5 By applying Theorem 1.9 with $u(x) = v(y) = 1$, $v_0 = 1$, $X = \bigcup_{i=1}^n X_i$, $X_i = [i-1, i)$ for $i = 1, 2, \ldots, n$, $d\mu(x) = dx$, $f(x,y) = f_i(y)$, and $f_0(x) = c_i$ for each $x \in X_i$, $i = 1, 2, \ldots, n$, we get the following version of (1.28):

$$\sum_{i=1}^n \left(c_i^p - \int_Y f_i^p(y) dv(y) \right)^{\frac{1}{p}} \leq \left(\left(\sum_{i=1}^n c_i \right)^p - \int_Y \left(\sum_{i=1}^n f_i(y) \right)^p dv(y) \right)^{\frac{1}{p}}$$

whenever

$$c_i \geq \left(\int_Y f_i^p(y) \, dv(y) \right)^{\frac{1}{p}}, \quad i = 1, 2, \ldots, n.$$

Remark 1.6 We can see from our proof that we can state a more general (refined) inequality by not using the finite estimate in (1.30).

We consider the functionals

$$B_1(v_0, v) = v_0 \left[\int_X f_0(x) u(x) \, d\mu(x) \right]^p$$
$$- \int_Y \left[\int_X f(x,y) u(x) \, d\mu(x) \right]^p v(y) \, dv(y)$$
$$- \left(\int_X \left[v_0 f_0^p(x) - \int_Y f^p(x,y) v(y) \, dv(y) \right]^{\frac{1}{p}} u(x) \, d\mu(x) \right)^p$$

and

$$B_2(v_0, v) = v_0 \left[\int_X f_0(x) u(x) \, d\mu(x) \right]^p$$

$$-\left(\int_X \left[\int_Y f^p(x,y)v(y)\,\mathrm{d}v(y)\right]^{\frac{1}{p}} u(x)\,\mathrm{d}\mu(x)\right)^p$$

$$-\left(\int_X \left[v_0 f_0^p(x) - \int_Y f^p(x,y)v(y)\,\mathrm{d}v(y)\right]^{\frac{1}{p}} u(x)\,\mathrm{d}\mu(x)\right)^p.$$

The proof of Theorem 1.9 shows that the inequality there can be replaced by the following (refined) version:

$$B_1(v_0, v) \geq B_2(v_0, v) \geq 0. \tag{1.31}$$

Example 1.6 Especially, by applying the previous inequality (1.31) for the special case pointed out in Example 1.5 we obtain the following refinement of the finite version of Bellman's inequality:

$$\left(\sum_{i=1}^n c_i\right)^p - \int_Y \left(\sum_{i=1}^n f_i(y)\right)^p \mathrm{d}v(y) - \left(\sum_{i=1}^n \left(c_i^p - \int_Y f_i^p(y)\,\mathrm{d}v(y)\right)^{\frac{1}{p}}\right)^p$$

$$\geq \left(\sum_{i=1}^n c_i\right)^p - \left(\sum_{i=1}^n \left(\int_Y f_i^p(y)\,\mathrm{d}v(y)\right)^{\frac{1}{p}}\right)^p$$

$$- \left(\sum_{i=1}^n \left(c_i^p - \int_Y f_i^p(y)\,\mathrm{d}v(y)\right)^{\frac{1}{p}}\right)^p \geq 0.$$

When $n = 2$, this is a refinement of (1.28) with $v_0 = 1$.

Finally, we point out a close relation between Theorems 1.7 and 1.9 (similar to that between Theorems 1.1 and 1.2), c.f. Theorem 1.3.

Example 1.7 Again as in Sect. 1.2.3 we put in (1.29) $f_0^{\frac{1}{p}}(x)$ and $f^{\frac{1}{p}}(x, y)$ instead of $f_0(x)$ and $f(x, y)$, respectively. By letting $p \to \infty$ and using the same power mean argument, we get inequality (1.23). In the discrete case a similar relation between the Bellman and Popoviciu inequalities was also pointed out in the paper [126] by S. Wu.

1.4 On the Gauss-Pólya Inequality

In [115, Vol. II] the following inequality involving decreasing function can be found: *If $f : [0, \infty) \to \mathbb{R}$ is a nonnegative nonincreasing function and if a and b are nonnegative real numbers, then*

1.4 On the Gauss-Pólya Inequality

$$\left(\int_0^\infty x^{a+b} f(x) \mathrm{d}x\right)^2$$
$$\leq \left(1 - \left(\frac{a-b}{a+b+1}\right)^2\right) \left(\int_0^\infty x^{2a} f(x) \mathrm{d}x\right) \left(\int_0^\infty x^{2b} f(x) \mathrm{d}x\right), \quad (1.32)$$

provided that all integrals exist.

If $a = 2$, $b = 0$ and using notation $m_r = \int_0^\infty x^r f(x) \mathrm{d}x$ with $m_0 = 1$, then inequality (1.32) becomes an inequality between the second and the fourth moments

$$5m_4 \geq 9m_2^2,$$

which was studied by C. F. Gauss in [44]. Nowadays, inequalities of that type are called the Gauss-Pólya inequalities. It is interesting that similar inequality holds if f is nondecreasing. Namely, already in the book [115], but in the first volume, we can find the result for a nondecreasing function f, which reads:

If $f : [0, 1] \to \mathbb{R}$ is a nonnegative nondecreasing function and if a and b are nonnegative real numbers, then

$$\left(\int_0^1 x^{a+b} f(x) \mathrm{d}x\right)^2$$
$$\geq \left(1 - \left(\frac{a-b}{a+b+1}\right)^2\right) \left(\int_0^1 x^{2a} f(x) \mathrm{d}x\right) \left(\int_0^1 x^{2b} f(x) \mathrm{d}x\right), \quad (1.33)$$

provided that all integrals exist.

During the last century, these inequalities were generalized in different ways (see for example [15, 83, 124, 125] and references therein). Here we point out a result involving derivatives of functions and which is based on a unified treatment of the both inequalities. It is published in [109].

Theorem 1.10 *Let $p_1, \ldots, p_n$ be positive real numbers such that $\sum_{i=1}^n \frac{1}{p_i} = 1$. Let $f, x_1, \ldots, x_n : [c, d] \to \mathbb{R}$ be nonnegative nondecreasing functions such that $x_1, \ldots, x_n$ and $\prod_{i=1}^n x_i^{1/p_i}$ have continuous first derivatives. Then*

$$\int_c^d \left(\prod_{i=1}^n (x_i(t))^{1/p_i}\right)' f(t) \mathrm{d}t \geq \prod_{i=1}^n \left(\int_c^d x_i'(t) f(t) \mathrm{d}t\right)^{1/p_i}. \quad (1.34)$$

If f is a nonincreasing nonnegative function, $x_1, \ldots, x_n$ additionally satisfy $x_i(c) = 0$, $i = 1, \ldots, n$, then the reversed inequality (1.34) holds.

Example 1.8 It is obvious that for $n = 2$, $p_1 = p_2 = 2$, $x_1(t) = t^{2a+1}$, and $x_2(t) = t^{2b+1}$, inequality (1.34) coincides with inequality (1.33) with $c = 0$, $d = 1$. Moreover, with $c = 0$ and letting $d \to \infty$, we see that reversed inequality (1.34) holds. Hence, Theorem 1.10 is a genuine generalization of both Gauss-Pólya inequalities (1.32) and (1.33) mentioned before.

Next we state and prove a continuous form of the Gauss-Pólya inequality involving derivatives (see [92]).

Theorem 1.11

(i) *Suppose that $f : [a, b] \to \mathbb{R}$ is nonnegative and nondecreasing, and $g(x, t)$ is a positive measurable function on $X \times [a, b]$ such that the functions $t \mapsto g(x, t)$ (for $x \in X$) are nondecreasing with a continuous first derivative. If $p(x) > 0$, $\int_X \frac{d\mu(x)}{p(x)} = 1$, then*

$$\int_a^b \left[\exp\left(\int_X \log g(x, t) \frac{d\mu(x)}{p(x)} \right) \right]' f(t) \, dt$$
$$\geq \exp\left[\int_X \log\left(\int_a^b g_t'(x, t) f(t) \, dt \right) \frac{d\mu(x)}{p(x)} \right], \quad (1.35)$$

provided that all integrals exist and where $g_t'(x, t) = \frac{d}{dt} g(x, t)$.

(ii) *Suppose that $f : [a, b] \to \mathbb{R}$ is nonnegative and nonincreasing, and $g(x, t)$ is a positive measurable function on $X \times [a, b]$ with respect to measure $\mu \times (-f) dx$ such that the functions $t \mapsto g(x, t)$ (for $x \in X$) are nondecreasing with a continuous first derivative and $g(x, a) = 0$ for all $x \in X$. If $p(x) > 0$, $\int_X \frac{d\mu(x)}{p(x)} = 1$, then the reverse inequality in (1.35) holds.*

Proof

(i) We assume that $f(b) > 0$, else $f \equiv 0$, and we get zero in both of the sides of inequality.

Integrating by parts and then using the continuous Hölder inequality (Theorem 1.1), we get

$$\int_a^b \left[\exp\left(\int_X \log g(x, t) \frac{d\mu(x)}{p(x)} \right) \right]' f(t) \, dt$$
$$= f(b) \exp\left(\int_X \log g(x, b) \frac{d\mu(x)}{p(x)} \right) - f(a) \exp\left(\int_X \log g(x, a) \frac{d\mu(x)}{p(x)} \right)$$

1.4 On the Gauss-Pólya Inequality

$$-\int_a^b \exp\left(\int_X \log g(x,t) \frac{d\mu(x)}{p(x)}\right) df(t)$$

$$\geq f(b) \exp\left(\int_X \log g(x,b) \frac{d\mu(x)}{p(x)}\right) - f(a) \exp\left(\int_X \log g(x,a) \frac{d\mu(x)}{p(x)}\right)$$

$$- \exp\left[\int_X \log\left(\int_a^b g(x,t) df(t)\right) \frac{d\mu(x)}{p(x)}\right].$$

Using Corollary 1.2 with

$$m=3, \quad w_1 = f(b) > 0, \quad w_2 = f(a), \quad w_3 = 1, \quad a_1(x) = (g(x,b))^{\frac{1}{p(x)}},$$

$$a_2(x) = (g(x,a))^{\frac{1}{p(x)}} \quad \text{and} \quad a_3(x) = \left(\int_a^b g(x,t) df(t)\right)^{\frac{1}{p(x)}},$$

we obtain that

$$f(b) \exp\left(\int_X \log g(x,b) \frac{d\mu(x)}{p(x)}\right) - f(a) \exp\left(\int_X \log g(x,a) \frac{d\mu(x)}{p(x)}\right)$$

$$- \exp\left[\int_X \log\left(\int_a^b g(x,t) df(t)\right) \frac{d\mu(x)}{p(x)}\right]$$

$$\geq \exp\left[\int_X \log\left(f(b)g(x,b) - f(a)g(x,a) - \int_a^b g(x,t) df(t)\right) \frac{d\mu(x)}{p(x)}\right]$$

$$= \exp\left[\int_X \log\left(\int_a^b f(t) g'_t(x,t) dt\right) \frac{d\mu(x)}{p(x)}\right],$$

so (1.35) holds. Here we integrated by parts once more. The proof of (i) is complete.

In the proof of (ii) we can use the same method as in the proof of (i) only instead of Corollary 1.2 we use Corollary 1.1 with

$$m=2, \quad w_1 = f(b), \quad w_2 = 1,$$

$$a_1(x) = (g(x,b))^{\frac{1}{p(x)}} \quad \text{and} \quad a_2(x) = \left(\int_a^b g(x,t) d(-f)(t)\right)^{\frac{1}{p(x)}},$$

so we omit the details. The proof is complete. □

Example 1.9 If $X = [0,n)$, $p(x) = p_i$, $x \in [i-1, i)$, $g(x,t) = x_i(t)$, $x \in [i-1, i)$, then Theorem 1.11 becomes Theorem 1.10.

If in Theorem 1.11 we put $g(x, t) = (g(t))^{a(x)p(x)+1}$, then we get the following result, which is a continuous form of the Gauss-Pólya inequality proved in [125].

Theorem 1.12 *Let $g : [a, b] \to \mathbb{R}$ be a nonnegative increasing differentiable function and let $f : [a, b] \to \mathbb{R}$ be a nonnegative nondecreasing function. Let p be a positive function such that $\int_X \frac{1}{p(x)} d\mu(x) = 1$. If $a(x)$ satisfies that $a(x) > -1/p(x)$, then*

$$\int_a^b (g(t))^{\int_X a(x) d\mu(x)} g'(t) f(t) \, dt \geq \frac{\exp\left(\int_X \log[a(x)p(x) + 1] \frac{d\mu(x)}{p(x)}\right)}{1 + \int_X a(x) d\mu(x)} \times$$

$$\times \exp\left\{\int_X \left[\log\left(\int_a^b f(t) g'(t) (g(t))^{a(x)p(x)} \, dt\right)\right] \frac{d\mu(x)}{p(x)}\right\},$$

provided that all integrals exist.

The following theorem is also a generalization of the Gauss-Pólya inequality. A particular case of it involving integrals was given in [2].

Theorem 1.13 *Let $w, w_x, (x \in X)$ be nonnegative and integrable functions on $[a, b]$ such that $\int_a^b w_x(t) dt \neq 0$, $\int_a^b w(t) dt \neq 0$, and let $W_x(t)$ and W be defined by*

$$W_x(t) = \frac{\int_a^t w_x(s) \, ds}{\int_a^b w_x(s) \, ds}, \quad W(t) = \frac{\int_a^t w(s) \, ds}{\int_a^b w(s) \, ds}.$$

Let $p(x) > 0$ and $\int_X \frac{dx}{p(x)} = 1$.

(a) If f is a nonnegative nonincreasing function on $[a, b]$ and if

$$\exp \int_X \log W_x(t) \frac{dx}{p(x)} \geq W(t) \tag{1.36}$$

for all $t \in [a, b]$, then

$$\frac{\int_a^b w(t) f(t) \, dt}{\int_a^b w(t) \, dt} \leq \exp\left[\int_X \log\left(\frac{\int_a^b w_x(t) f(t) \, dt}{\int_a^b w_x(t) \, dt}\right) \frac{1}{p(x)} dx\right]. \tag{1.37}$$

(b) If f is a nonnegative nondecreasing function on $[a, b]$ and the reverse inequality in (1.36) holds, then inequality (1.37) is reversed.

1.4 On the Gauss-Pólya Inequality

Proof

(a) Let us denote the right-hand side of (1.37) by A. Using integration by parts, we find that

$$A = \exp\left[\int_X \log\left(\int_a^b \frac{d}{dt} W_x(t) f(t)\, dt\right) \frac{dx}{p(x)}\right]$$

$$= \exp\left[\int_X \log\left(f(b) + \int_a^b W_x(t)\, d(-f(t))\, dt\right) \frac{dx}{p(x)}\right].$$

Putting in Corollary 1.1 the following: $m = 2$, $w_1 = w_2 = 1$, $u(x) = 1$, $d\mu(x) = dx$, $a_1 = (f(b))^{\frac{1}{p(x)}}$ and $a_2 = \left(\int_a^b W_x(t)\, d(-f(t))\right)^{\frac{1}{p(x)}}$, we get

$$A \geq f(b) + \exp\left[\int_X \log\left(\int_a^b W_x(t)\, d(-f(t))\right) \frac{dx}{p(x)}\right].$$

Once again we use the continuous Hölder inequality (Theorem 1.1) and get that

$$A \geq f(b) + \int_a^b \exp\left[\int_X \log(W_x(t)) \frac{dx}{p(x)}\right] d(-f(t))$$

$$\geq f(b) + \int_a^b W(t)\, d(-f(t)) = \int_a^b W'(t) f(t)\, dt = \frac{\int_a^b w(t) f(t)\, dt}{\int_a^b w(t)\, dt}.$$

(b) The proof is similar, and only instead of the discrete Hölder inequality we use the Popoviciu inequality from Corollary 1.2. So we omit the details and the proof is complete. □

Remark 1.7 Putting

$$w_x(t) = g'_t(x, t) \quad \text{and} \quad w(t) = \left(\exp\left(\int_X \log g(x, t) \frac{dx}{p(x)}\right)\right)',$$

with $g(x, a) = 0$, then we get

$$W_x(t) = \frac{g(x, t)}{g(x, b)} \quad \text{and} \quad W(t) = \frac{\exp\left(\int_X \log g(x, t) \frac{dx}{p(x)}\right)}{\exp\left(\int_X \log g(x, b) \frac{dx}{p(x)}\right)}$$

and, hence,

$$\exp \int_X \log W_x(t) \frac{\mathrm{d}x}{p(x)} = W(t).$$

Therefore, by applying Theorem 1.13 in both cases (a) and (b), we rediscover the result of Theorem 1.11.

Proof Let us suppose that f is nondecreasing. We conclude that

$$\int_a^b \left[\exp\left(\int_X \log g(x,t) \frac{\mathrm{d}x}{p(x)} \right) \right]' f(t)\,\mathrm{d}t = \frac{\int_a^b w(t)f(t)\,\mathrm{d}t}{\int_a^b w(t)\,\mathrm{d}t} \cdot \int_a^b w(t)\,\mathrm{d}t$$

$$\geq \exp\left[\int_X \log\left(\frac{\int_a^b w_x(t)f(t)\,\mathrm{d}t}{\int_a^b w_x(t)\,\mathrm{d}t} \right) \frac{\mathrm{d}x}{p(x)} \right] \times$$

$$\times \exp\left(\int_X \log(g(x,b) - g(x,a)) \frac{\mathrm{d}x}{p(x)} \right)$$

$$= \exp\left[\int_X \log\left(\int_a^b g'_t(x,t)f(t)\,\mathrm{d}t \right) \frac{\mathrm{d}x}{p(x)} \right]$$

and the proof is complete. □

Refinements of Continuous Forms of Inequalities

A motivation for the study described in this chapter can be found in several papers devoted to refining of different inequalities which appeared in the last decade. One very representative example is the refinement related to the Hölder inequality. Here we present the mentioned refinement that is created by combining the results from [50] and [51] (see also [54]).

Theorem 2.1 *Let $p, q > 1$ be such that $\frac{1}{p} + \frac{1}{q} = 1$, f and g be real functions defined on $[a, b]$ such that $|f|^p$ and $|g|^q$ are integrable on $[a, b]$, $-\infty \leq a < b \leq \infty$ and α_i, $i = 1, 2, \ldots, n$, are nonnegative continuous functions on $[a, b]$ such that $\sum_{i=1}^{n} \alpha_i(t) = 1$ for all $t \in [a, b]$. Then*

$$\int_a^b |f(y)g(y)|\,dy \leq \sum_{i=1}^{n} \left[\int_a^b \alpha_i(y)|f(y)|^p\,dy\right]^{\frac{1}{p}} \left[\int_a^b \alpha_i(y)|g(y)|^q\,dy\right]^{\frac{1}{q}}$$

$$\leq \left(\int_a^b |f(y)|^p\,dy\right)^{\frac{1}{p}} \left(\int_a^b |g(y)|^q\,dy\right)^{\frac{1}{q}}. \quad (2.1)$$

In [50], an interesting application of the previous result (2.1) connected with the Hermite-Hadamard inequality was given. Namely, for $n = 2$, $\alpha_1(t) = 1 - t$, and $\alpha_2(t) = t$ the following refinement of the known estimation for a difference between two sides from the Hermite-Hadamard inequality was obtained.

Theorem 2.2 *Let $p, q > 1$ be such that $\frac{1}{p} + \frac{1}{q} = 1$. Let $f : I^o \subseteq \mathbb{R} \to \mathbb{R}$ be a differentiable mapping, and let $a, b \in I^o$ with $a < b$. If the mapping $|f'|^q$ is convex on*

$[a, b]$, then

$$\left| \frac{f(a) + f(b)}{2} - \frac{1}{b-a} \int_a^b f(x)\,dx \right|$$
$$\leq \frac{b-a}{4(p+1)^{1/p}} \left[\left(\frac{2|f'(a)|^q + |f'(b)|^q}{3} \right)^{1/q} + \left(\frac{|f'(a)|^q + 2|f'(b)|^q}{3} \right)^{1/q} \right]$$
$$\leq \frac{b-a}{2(p+1)^{1/p}} \left(\frac{|f'(a)|^q + |f'(b)|^q}{2} \right)^{1/q}.$$

In this chapter we give a unified treatment of the mentioned study deriving refinements of some continuous forms of the Hölder and other classical inequalities. Our efforts lead to rediscovering known results but also to obtain several new refinements. Results presented in this chapter are mainly published in [101], see also [7].

2.1 Refinements of the Hölder and Popoviciu Inequalities

First we present the following refinement of the continuous form of the Hölder inequality.

Theorem 2.3 *Let $f(x, y)$ be a positive and measurable function on $(X \times Y, \mu \times \nu)$ and let $u(x)$ and $v(y)$ be weight functions on X and Y, respectively, such that $\int_X u(x)\,d\mu(x) = 1$ and $\int_Y f(x, y)v(y)d\nu(y) > 0$ μ-a.e. Moreover, let (Z, dz) be a measure space and $\alpha(z, y)$ be a nonnegative integrable function on $Z \times Y$ such that $\int_Y \alpha(z, y) f(x, y)v(y)d\nu(y) > 0$ μ-a.e and*

$$\int_Z \alpha(z, y)\,dz = 1, \qquad \text{for } y \in Y. \tag{2.2}$$

Then the following refinement of continuous form (1.4) *of the Hölder inequality holds:*

$$\int_Y \exp\left(\int_X \log f(x, y) u(x)\,d\mu(x) \right) v(y)\,d\nu(y)$$
$$\leq \int_Z \left[\exp \int_X \log \left(\int_Y \alpha(z, y) f(x, y) v(y)\,d\nu(y) \right) u(x)\,d\mu(x) \right] dz$$
$$\leq \exp \left[\int_X \log \left(\int_Y f(x, y)v(y)d\nu(y) \right) u(x)\,d\mu(x) \right]. \tag{2.3}$$

Proof The proof is based on the use of the Fubini theorem and the continuous form of the Hölder inequality. Namely, by using condition (2.2) and the Fubini theorem, we get

2.1 Refinements of the Hölder and Popoviciu Inequalities

$$\int_Y \exp\left(\int_X \log f(x,y) u(x) \, d\mu(x)\right) v(y) \, d\nu(y)$$

$$= \int_Y \left[\int_Z \alpha(z,y) \exp\left(\int_X \log f(x,y) u(x) \, d\mu(x)\right) dz\right] v(y) \, d\nu(y)$$

$$= \int_Z \left[\int_Y \exp\left(\int_X \log f(x,y) u(x) \, d\mu(x)\right) \alpha(z,y) v(y) \, d\nu(y)\right] dz$$

$$\leq \int_Z \exp\left[\int_X \log\left(\int_Y f(x,y) \alpha(z,y) v(y) \, d\nu(y)\right) u(x) \, d\mu(x)\right] dz,$$

where in the last inequality we apply (1.4). Hence, the first inequality in (2.3) is proved. By using (1.4) again, we obtain that

$$\int_Z \exp\left[\int_X \log\left(\int_Y f(x,y) \alpha(z,y) v(y) \, d\nu(y)\right) u(x) \, d\mu(x)\right] dz$$

$$\leq \exp\left[\int_X \log\left(\int_Z \left(\int_Y f(x,y) \alpha(z,y) v(y) \, d\nu(y)\right) dz\right) u(x) \, d\mu(x)\right]$$

$$= \exp\left[\int_X \log\left(\int_Y \left(\int_Z f(x,y) \alpha(z,y) \, dz\right) v(y) \, d\nu(y)\right) u(x) \, d\mu(x)\right]$$

$$= \exp\left[\int_X \log\left(\int_Y f(x,y) v(y) \, d\nu(y)\right) u(x) \, d\mu(x)\right],$$

where the Fubini theorem and (2.2) are used in the last two equalities. Thus, (2.3) holds and the proof is complete. □

The immediate consequences of Theorem 2.3 are given in the following Corollary.

Corollary 2.1 *(a) Let the assumptions of Theorem 2.3 hold and let $p(x)$ be a measurable function on X. Then the following refinement of the continuous Hölder inequality holds:*

$$\int_Y \exp\left(\int_X p(x) \log f(x,y) \, u(x) \, d\mu(x)\right) v(y) \, d\nu(y)$$

$$\leq \int_Z \left[\exp \int_X \left(\int_Y \alpha(z,y) f(x,y)^{p(x)} v(y) \, d\nu(y)\right) u(x) \, d\mu(x)\right] dz$$

$$\leq \exp\left[\int_X \log\left(\int_Y f(x,y)^{p(x)} v(y) \, d\nu(y)\right) u(x) \, d\mu(x)\right]. \quad (2.4)$$

(b) Let $p, q > 1$ be such that $\dfrac{1}{p} + \dfrac{1}{q} = 1$. If $f(y)$, $g(y)$, $\alpha(z,y)$ are nonnegative functions such that $fgv \in L_1(Y)$, $fv, \alpha^{1/p}(z,.)fv \in L_p(Y)$, $gv, \alpha^{1/q}(y,.)gv \in L_q(Y)$, and

$\int_Z \alpha(z, y)\, dz = 1$ *for all* $y \in Y$, *then the following refinement of the Hölder inequality holds:*

$$\|fg\|_{1,v} \leq \int_Z \|\alpha^{1/p}(z,.)f(.)\|_{p,v} \cdot \|\alpha^{1/q}(z,.)g(.)\|_{q,v}\, dz \leq \|f\|_{p,v}\|g\|_{q,v}, \quad (2.5)$$

where $\|f\|_{p,v} = \left(\int_Y f^p(y) v(y)\, dv(y)\right)^{1/p}$.

(c) *Let* $p, q > 1$ *be such that* $\dfrac{1}{p} + \dfrac{1}{q} = 1$. *If* f, g, α *are nonnegative functions on* Y *such that* $fgv \in L_1(Y)$, $fv, \alpha^{1/p} fv \in L_p(Y)$, $gv, \alpha^{1/q} gv \in L_q(Y)$ *and* $\alpha(y) \leq 1$ *for all* $y \in Y$, *then we find that also the following refinement of Hölder inequality holds:*

$$\|fg\|_{1,v} \leq \|\alpha^{\frac{1}{p}} f\|_{p,v} \cdot \|\alpha^{\frac{1}{q}} g\|_{q,v} + \|(1-\alpha)^{\frac{1}{p}} f\|_{p,v} \cdot \|(1-\alpha)^{\frac{1}{q}} g\|_{q,v}$$
$$\leq \|f\|_{p,v}\|g\|_{q,v}. \quad (2.6)$$

Proof (a) Inequality (2.4) is a simple consequence of Theorem 2.3 applied with $f(x, y)^{p(x)}$ in place of $f(x, y)$.

(b) By putting in the (a) part of this corollary: $X = X_1 \cup X_2$, $X_1 \cap X_2 = \emptyset$, such that $\int_{X_1} u(x)\, d\mu(x) = \dfrac{1}{p}$, $\int_{X_2} u(x)\, d\mu(x) = \dfrac{1}{q}$, and

$$f(x, y) = \begin{cases} f(y), & x \in X_1 \\ g(y), & x \in X_2, \end{cases} \qquad p(x) = \begin{cases} p, & x \in X_1 \\ q, & x \in X_2, \end{cases}$$

we get inequality (2.5).

(c) Inequality (2.6) follows from inequality (2.5) by taking:

$$Z = [0, 2],\ Z_1 = [0, 1),\ Z_2 = [1, 2]\ \text{and}\ \alpha(z, y) = \begin{cases} \alpha(y), & z \in Z_1 \\ 1 - \alpha(y), & z \in Z_2. \end{cases}$$

The proof is complete. □

A comparison between Corollary 2.1 and the known results is given in the following remark. In fact, Corollary 2.1 is a generalization of the results from [50] and [54].

Remark 2.1

(a) The chain of inequalities from part (c) of the above corollary for $Y = [a, b]$ and $v(y)\, dv(y) = dy$ can be found in [54], while the first inequality was proved in [50]. Moreover, if $Y = [a, b]$, $v(y)\, dv(y) = dy$, and $\alpha(t) = b - t$, the chain of inequalities from part (c) was proved in [50].

2.1 Refinements of the Hölder and Popoviciu Inequalities

(b) By using the same idea as in part b) we can derive a refinement of the Hölder inequality with n functions involved ($n = 2, 3, \ldots$).

In Chap. 1, Sect. 1.2.1 (see (1.6) and see also [38, VI.11.35]), the following interesting inequality was derived:

$$\exp\left[\int_X \log a(x) u(x) \, d\mu(x)\right] + \exp\left[\int_X \log b(x) u(x) \, d\mu(x)\right]$$
$$\leq \exp\left[\int_X \log (a(x) + b(x)) u(x) \, d\mu(x)\right], \qquad (2.7)$$

provided that all integrals exist and $a(.)$ and $b(.)$ are positive.

Next theorem consists of a refinement of inequality (2.7).

Theorem 2.4 *Let (X, μ), (Y, ν) and (Z, dz) be σ-finite measure spaces. Let a and b be positive measurable functions on X, $u(x)$ be a weight on X, $v(y)$ be a weight on Y, and $\alpha(z, y)$ be a nonnegative function on $Z \times Y$ such that $\int_Z \alpha(z, y) \, dz = 1$ for all $y \in Y$. If Y has a partition $Y = Y_1 \cup Y_2$, such that $\int_{Y_i} v(y) \, d\nu(y) = 1$, $i = 1, 2$, and the integrals $A(z) = \int_{Y_1} \alpha(z, y) v(y) \, d\nu(y)$ and $B(z) = \int_{Y_2} \alpha(z, y) v(y) \, d\nu(y)$ exist, then*

$$\exp\left[\int_X \log a(x) u(x) \, d\mu(x)\right] + \exp\left[\int_X \log b(x) u(x) \, d\mu(x)\right]$$
$$\leq \int_Z \left[\exp \int_X \log \Big(A(z) a(x) + B(z) b(x)\Big) u(x) \, d\mu(x)\right] dz$$
$$\leq \exp\left[\int_X \log \Big(a(x) + b(x)\Big) u(x) \, d\mu(x)\right]. \qquad (2.8)$$

Proof We apply (2.3) in Theorem 2.3 with

$$f(x, y) = \begin{cases} a(x), & y \in Y_1 \\ b(x), & y \in Y_2 \end{cases}$$

and obtain (2.8). The proof is complete. □

Remark 2.2 Let us take

$$Z = [0, 2], \quad Z_1 = [0, 1), \quad Z_2 = [1, 2],$$

and let Y have a partition $Y = Y_1 \cup Y_2$ such that $\int_{Y_i} v(y) dv(y) = 1, i = 1, 2$.
Denote

$$\alpha(z, y) = \begin{cases} A, & z \in Z_1, \ y \in Y_1 \\ 1 - A, & z \in Z_1, \ y \in Y_2 \\ B, & z \in Z_2, \ y \in Y_1 \\ 1 - B, & z \in Z_2, \ y \in Y_2, \end{cases}$$

where $A, B \in [0, 1]$. Obviously, the condition on $\alpha(z, y)$ is fulfilled. In this case, the middle term in (2.8) takes the form

$$\exp\left[\int_X \log\Big(Aa(x) + Bb(x)\Big) u(x) d\mu(x)\right]$$
$$+ \exp\left[\int_X \log\Big((1-A)a(x) + (1-B)b(x)\Big) u(x) d\mu(x)\right],$$

where $A, B \in [0, 1]$.

Let us recall that the continuous form of the Popoviciu inequality was given in Sect. 1.3.1 (see (1.23)). The refinement of it is given as the following result:

Theorem 2.5 *Let $u(x)$ and $v(y)$ be weight functions on the measure spaces (X, μ) and (Y, v), respectively, where $\int_X u(x) d\mu(x) = 1$. Let $f(x, y)$ be a positive measurable function on $X \times Y$, $v_0 > 0$, and assume that $f_0(x)$ is a function on X such that $v_0 f_0(x) > \int_Y f(x, y) v(y) dv(y)$, for all $x \in X$. Moreover, let $\alpha(z, y)$ be a nonnegative integrable function on $Z \times Y$ such that $\int_Z \alpha(z, y) dz = 1$ for $y \in Y$, where (Z, dz) is a σ-finite measure space. Then the following refinement of the continuous form of the Popoviciu inequality (1.23) holds:*

$$\exp\left(\int_X \log\Big(v_0 f_0(x)\Big) u(x) d\mu(x)\right)$$
$$- \int_Y \exp\left(\int_X \log(f(x, y)) u(x) d\mu(x)\right) v(y) dv(y)$$
$$\geq \exp\left(\int_X \log\Big(v_0 f_0(x)\Big) u(x) d\mu(x)\right)$$
$$- \int_Z \left[\exp \int_X \log\left(\int_Y f(x, y) \alpha(z, y) v(y) dv(y)\right) u(x) d\mu(x)\right] dz$$
$$\geq \exp\left[\int_X \log\Big(v_0 f_0(x) - \int_Y f(x, y) v(y) dv(y)\Big) u(x) d\mu(x)\right] \geq 0. \quad (2.9)$$

2.1 Refinements of the Hölder and Popoviciu Inequalities

Proof From Theorem 2.3 we get

$$-\int_Y \exp\left(\int_X \log f(x,y) u(x)\, d\mu(x)\right) v(y)\, d\nu(y)$$
$$\geq -\int_Z \left[\exp \int_X \log\left(\int_Y \alpha(z,y) f(x,y) v(y)\, d\nu(y)\right) u(x)\, d\mu(x)\right] dz$$
$$\geq -\exp\left[\int_X \log\left(\int_Y f(x,y) v(y) d\nu(y)\right) u(x)\, d\mu(x)\right].$$

Adding to the each side of the above inequality expression

$$\exp\left(\int_X \log\left(v_0 f_0(x)\right) u(x)\, d\mu(x)\right)$$
$$-\exp\left[\int_X \log\left(v_0 f_0(x) - \int_Y f(x,y) v(y)\, d\nu(y)\right) u(x)\, d\mu(x)\right],$$

we obtain

$$\exp\left(\int_X \log\left(v_0 f_0(x)\right) u(x)\, d\mu(x)\right)$$
$$-\exp\left[\int_X \log\left(v_0 f_0(x) - \int_Y f(x,y) v(y)\, d\nu(y)\right) u(x)\, d\mu(x)\right]$$
$$-\int_Y \exp\left(\int_X \log f(x,y) u(x)\, d\mu(x)\right) v(y)\, d\nu(y)$$
$$\geq \exp\left(\int_X \log\left(v_0 f_0(x)\right) u(x)\, d\mu(x)\right)$$
$$-\exp\left[\int_X \log\left(v_0 f_0(x) - \int_Y f(x,y) v(y)\, d\nu(y)\right) u(x)\, d\mu(x)\right]$$
$$-\int_Z \left[\exp \int_X \log\left(\int_Y \alpha(z,y) f(x,y) v(y)\, d\nu(y)\right) u(x)\, d\mu(x)\right] dz$$
$$\geq \exp\left(\int_X \log\left(v_0 f_0(x)\right) u(x)\, d\mu(x)\right)$$
$$-\exp\left[\int_X \log\left(v_0 f_0(x) - \int_Y f(x,y) v(y)\, d\nu(y)\right) u(x)\, d\mu(x)\right]$$
$$-\exp\left[\int_X \log\left(\int_Y f(x,y) v(y) d\nu(y)\right) u(x)\, d\mu(x)\right]. \tag{2.10}$$

The right-hand side of (2.10) is equal to functional $P_2(v_0, v)$ for which nonnegativity is already proved, see (1.26) in Sect. 1.3.1. Hence, we have

$$\exp\left(\int_X \log\left(v_0 f_0(x)\right) u(x) \, d\mu(x)\right)$$

$$- \exp\left[\int_X \log\left(v_0 f_0(x) - \int_Y f(x, y) v(y) \, dv(y)\right) u(x) \, d\mu(x)\right]$$

$$- \int_Y \exp\left(\int_X \log f(x, y) u(x) \, d\mu(x)\right) v(y) \, dv(y)$$

$$\geq \exp\left(\int_X \log\left(v_0 f_0(x)\right) u(x) \, d\mu(x)\right)$$

$$- \exp\left[\int_X \log\left(v_0 f_0(x) - \int_Y f(x, y) v(y) \, dv(y)\right) u(x) \, d\mu(x)\right]$$

$$- \int_Z \left[\exp \int_X \log\left(\int_Y \alpha(z, y) f(x, y) v(y) \, dv(y)\right) u(x) \, d\mu(x)\right] dz \geq 0$$

and by now adding to each side of this inequality the term

$$\exp\left[\int_X \log\left(v_0 f_0(x) - \int_Y f(x, y) v(y) \, dv(y)\right) u(x) \, d\mu(x)\right],$$

we obtain (2.9). The proof is complete. □

In the next remark we will point out the fact that in this case our result is new even for the case with only two functions involved.

Remark 2.3 Let $u(x) = 1$, $v_0 = 1$, $X = X_1 \cup X_2$, $X_1 \cap X_2 = \emptyset$ with

$$\int_{X_1} d\mu(x) = \frac{1}{p}, \int_{X_2} d\mu(x) = \frac{1}{q}, \text{ where } \frac{1}{p} + \frac{1}{q} = 1;$$

$f_0(x) = c_1^p$, $f(x, y) = f^p(y)$ for each $x \in X_1$ and $f_0(x) = c_2^q$, $f(x, y) = g^q(y)$ for each $x \in X_2$.

Then inequality (2.9) becomes

$$c_1 c_2 - \|fg\|_{1,v} \geq c_1 c_2 - \int_Z \|\alpha^{\frac{1}{p}}(z, .) f(.)\|_{p,v} \|\alpha^{\frac{1}{q}}(z, .) g(.)\|_{q,v} \, dz$$

$$\geq (c_1^p - \|f\|_{p,v}^p)^{\frac{1}{p}} \cdot (c_2^q - \|g\|_{q,v}^q)^{\frac{1}{q}}$$

and this is a continuous refinement of the integral Popoviciu inequality (1.21) for two functions.

Moreover, as in the part (c) of Corollary 2.1 for α, $0 \leq \alpha(y) \leq 1$ on Y, we get that

$$c_1 c_2 - \|fg\|_{1,v}$$
$$\geq c_1 c_2 - \left(\|\alpha^{\frac{1}{p}} f\|_{p,v} \cdot \|\alpha^{\frac{1}{q}} g\|_{q,v} + \|(1-\alpha)^{\frac{1}{p}} f\|_{p,v} \cdot \|(1-\alpha)^{\frac{1}{q}} g\|_{q,v}\right)$$
$$\geq \left(c_1^p - \|f\|_{p,v}^p\right)^{\frac{1}{p}} \cdot \left(c_2^q - \|g\|_{q,v}^q\right)^{\frac{1}{q}},$$

where $\dfrac{1}{p} + \dfrac{1}{q} = 1$, $p, q > 1$.

2.2 Refinements of the Minkowski and Bellman Inequalities

In Sects. 1.2.2 and 1.3.2 we stated continuous forms of the Minkowski and Bellman inequalities. In this section we state and prove their refinements. Our first main result in this section is the following refinement of the continuous Minkowski inequality (1.10):

Theorem 2.6 *Let $f(x, y)$ be a nonnegative and measurable function on $(X \times Y, \mu \times \nu)$, and let $u(x)$ and $v(y)$ be weight functions on X and Y, respectively. Moreover, let $\alpha(z, y)$ be a nonnegative integrable function on $Z \times Y$ such that $\int_Z \alpha(z, y)\, dz = 1$ for $y \in Y$, where (Z, dz) is a σ-finite measure space. If $p \geq 1$, then*

$$\int_Y \left(\int_X f(x, y) u(x)\, d\mu(x)\right)^p v(y)\, d\nu(y)$$
$$\leq \int_Z \left[\int_X \left(\int_Y \alpha(z, y) f^p(x, y) v(y)\, d\nu(y)\right)^{1/p} u(x)\, d\mu(x)\right]^p dz$$
$$\leq \left[\int_X \left(\int_Y f^p(x, y) v(y)\, d\nu(y)\right)^{1/p} u(x)\, d\mu(x)\right]^p. \quad (2.11)$$

Proof By using the condition on the function $\alpha(z, y)$ and the Fubini theorem, we get

$$\int_Y \left(\int_X f(x, y) u(x)\, d\mu(x)\right)^p v(y)\, d\nu(y)$$
$$= \int_Y \left[\int_Z \alpha(z, y) \left(\int_X f(x, y) u(x)\, d\mu(x)\right)^p dz\right] v(y)\, d\nu(y)$$
$$= \int_Z \left[\int_Y \alpha(z, y) \left(\int_X f(x, y) u(x)\, d\mu(x)\right)^p v(y)\, d\nu(y)\right] dz.$$

Moreover, by using the continuous Minkowski inequality (1.10) on the term in the square brackets and, then, on the integrals over Z and X, we obtain that

$$\int_Z \left[\int_Y \alpha(z,y) \left(\int_X f(x,y) u(x) \, d\mu(x) \right)^p v(y) \, d\nu(y) \right] dz$$

$$\leq \int_Z \left[\int_X \left(\int_Y \alpha(z,y) f^p(x,y) v(y) \, d\nu(y) \right)^{1/p} u(x) \, d\mu(x) \right]^p dz$$

$$\leq \left[\int_X \left[\int_Z \left(\int_Y \alpha(z,y) f^p(x,y) v(y) \, d\nu(y) \right) dz \right]^{1/p} u(x) \, d\mu(x) \right]^p$$

$$= \left[\int_X \left(\int_Y \left(\int_Z \alpha(z,y) f^p(x,y) \, dz \right) v(y) \, d\nu(y) \right)^{1/p} u(x) \, d\mu(x) \right]^p$$

$$= \left[\int_X \left(\int_Y f^p(x,y) v(y) \, d\nu(y) \right)^{1/p} u(x) \, d\mu(x) \right]^p,$$

where we also used the Fubini theorem and the condition on the function $\alpha(z,y)$. The proof is complete. □

Example 2.1 Putting

$$f(x,y) = \begin{cases} \dfrac{f(y)}{\alpha_1}, & x \in X_1 \\ \dfrac{g(y)}{\alpha_2}, & x \in X_2, \end{cases}$$

where $X = X_1 \cup X_2$, $X_1 \cap X_2 = \emptyset$, $\int_{X_1} u(x) \, d\mu(x) = \alpha_1$, and $\int_{X_2} u(x) \, d\mu(x) = \alpha_2$, $f, g > 0$, in (2.11) we get a refinement of usual integral Minkowski inequality with two functions involved. In particular, we have that

$$\|f+g\|_{p,v} \leq \left(\int_Z \left[\left(\int_Y \alpha(z,y) f^p(y) v(y) \, d\nu(y) \right)^{\frac{1}{p}} \right. \right.$$

$$\left. \left. + \left(\int_Y \alpha(z,y) g^p(y) v(y) \, d\nu(y) \right)^{\frac{1}{p}} \right]^p dz \right)^{\frac{1}{p}}$$

$$= \left(\int_Z \left(\|\alpha^{\frac{1}{p}}(z,.) f(.)\|_{p,v} + \|\alpha^{\frac{1}{p}}(z,.) g(.)\|_{p,v} \right)^p dz \right)^{\frac{1}{p}}$$

$$\leq \|f\|_{p,v} + \|g\|_{p,v},$$

which corresponds to the part (b) of Corollary 2.1.

2.2 Refinements of the Minkowski and Bellman Inequalities

Moreover, as in part (c) of Corollary 2.1 for α, $0 \leq \alpha(y) \leq 1$ on Y, we get that

$$\|f+g\|_{p,v} \leq \left[\left(\|\alpha^{\frac{1}{p}}f\|_{p,v} + \|\alpha^{\frac{1}{p}}g\|_{p,v}\right)^p + \left(\|(1-\alpha)^{\frac{1}{p}}f\|_{p,v} + \|(1-\alpha)^{\frac{1}{p}}g\|_{p,v}\right)^p\right]^{\frac{1}{p}}$$

$$\leq \|f\|_{p,v} + \|g\|_{p,v}.$$

It is clear that in the same way we can derive the corresponding refinements of the Minkowski inequality with n functions involved ($n = 2, 3, \ldots$).

Our refinement of the continuous form of the Bellman inequality (1.29) reads:

Theorem 2.7 *Let the assumptions of Theorem 2.6 hold. Assume that $v_0 > 0$ and $f_0(x)$ is a function on X such that $v_0 f_0^p(x) - \int_Y f^p(x,y) v(y) \, dv(y) \geq 0$. Let $\alpha(z,y)$ satisfy the assumptions of Theorem 2.5.*

Then the following refinement of the continuous Bellman inequality (1.29) holds for $p \geq 1$:

$$v_0 \left[\int_X f_0(x) u(x) \, d\mu(x)\right]^p - \int_Y \left(\int_X f(x,y) u(x) \, d\mu(x)\right)^p v(y) \, dv(y)$$

$$\geq v_0 \left[\int_X f_0(x) u(x) \, d\mu(x)\right]^p$$

$$- \int_Z \left[\int_X \left(\int_Y \alpha(z,y) f^p(x,y) v(y) \, dv(y)\right)^{1/p} u(x) \, d\mu(x)\right]^p dz$$

$$\geq \left(\int_X \left[v_0 f_0^p(x) - \int_Y f^p(x,y) v(y) \, dv(y)\right]^{\frac{1}{p}} u(x) \, d\mu(x)\right)^p. \quad (2.12)$$

Proof Multiplying inequality (2.11) by -1 and adding to each side of it the expression

$$v_0 \left[\int_X f_0(x) u(x) \, d\mu(x)\right]^p$$

$$- \left(\int_X \left[v_0 f_0^p(x) - \int_Y f^p(x,y) v(y) \, dv(y)\right]^{\frac{1}{p}} u(x) \, d\mu(x)\right)^p,$$

we find that:

$$v_0 \left[\int_X f_0(x) u(x) \, d\mu(x)\right]^p$$

$$- \left(\int_X \left[v_0 f_0^p(x) - \int_Y f^p(x,y) v(y) \, dv(y)\right]^{\frac{1}{p}} u(x) \, d\mu(x)\right)^p$$

$$-\int_Y \left(\int_X f(x,y)u(x)\,d\mu(x)\right)^p v(y)\,d\nu(y)$$

$$\geq v_0 \left[\int_X f_0(x)u(x)\,d\mu(x)\right]^p$$

$$-\left(\int_X \left[v_0 f_0^p(x) - \int_Y f^p(x,y)v(y)\,d\nu(y)\right]^{\frac{1}{p}} u(x)\,d\mu(x)\right)^p$$

$$-\int_Z \left[\int_X \left(\int_Y \alpha(z,y)f^p(x,y)v(y)\,d\nu(y)\right)^{1/p} u(x)\,d\mu(x)\right]^p dz$$

$$\geq v_0 \left[\int_X f_0(x)u(x)\,d\mu(x)\right]^p$$

$$-\left(\int_X \left[v_0 f_0^p(x) - \int_Y f^p(x,y)v(y)\,d\nu(y)\right]^{\frac{1}{p}} u(x)\,d\mu(x)\right)^p$$

$$-\left[\int_X \left(\int_Y f^p(x,y)v(y)\,d\nu(y)\right)^{1/p} u(x)\,d\mu(x)\right]^p.$$

The right-hand side of the above inequality is, in fact, equal to the functional B_2, defined in Sect. 1.2.2, and it is nonnegative (see (1.31)) and, hence, after simple transformation, we conclude that (2.12) holds. The proof is complete. □

It is always interesting to see how the general result is connected to the classical case (e.g., the case with only two functions involved).

Remark 2.4 In the case of two functions f and g we get the following "continuous" refinement of the Bellman inequality with $v_0 = 1$ (see (1.28)):

$$(c_1 + c_2)^p - \|f+g\|_{p,v}^p$$
$$\geq (c_1 + c_2)^p - \int_Z \left[\|\alpha^{\frac{1}{p}}(z,.)f(.)\|_{p,v} + \|\alpha^{\frac{1}{p}}(z,.)g(.)\|_{p,v}\right]^p dz$$
$$\geq \left[(c_1^p - \|f\|_{p,v}^p)^{\frac{1}{p}} + (c_2^p - \|g\|_{p,v}^p)^{\frac{1}{p}}\right]^p.$$

Besides the continuous form, it is instructive to see how that inequality looks like in an integral form. In particular, when instead of integral over Z we have only two summands, this inequality reads:

$$(c_1+c_2)^p - \|f+g\|_{p,v}^p \geq (c_1+c_2)^p - \left(\|\alpha^{\frac{1}{p}}f\|_{p,v} + \|\alpha^{\frac{1}{p}}g\|_{p,v}\right)^p$$
$$- \left(\|(1-\alpha)^{\frac{1}{p}}f\|_{p,v} + \|(1-\alpha)^{\frac{1}{p}}g\|_{p,v}\right)^p$$
$$\geq \left[(c_1^p - \|f\|_{p,v}^p)^{\frac{1}{p}} + (c_2^p - \|g\|_{p,v}^p)^{\frac{1}{p}}\right]^p,$$

where $0 \leq \alpha(y) \leq 1$.

2.3 Refinements of the Jensen Inequality and Its Reversed Version

In [117] the following continuous refinement of the Jensen inequality was given:

Theorem 2.8 *Let (X, μ) and (Z, λ) be two measure spaces, let w be a nonnegative measurable function on X, and let $\alpha : X \times Z \to [0, \infty)$ be a measurable function on $X \times Z$ satisfying*

$$\int_X \alpha(x,z)w(x)\,d\mu(x) = \int_X w(x)\,d\mu(x), \quad \text{for each } z \in Z \qquad (2.13)$$

and

$$\int_Z d\lambda(z) = 1, \quad \int_Z \alpha(x,z)\,d\lambda(z) = 1, \quad \text{for each } x \in X. \qquad (2.14)$$

If φ is a real convex function on the interval $I \subseteq \mathbb{R}$, then for the function $f : X \to I$, $wf, w(\varphi \circ f) \in L_1(X)$, it holds that

$$\varphi\left(\frac{1}{W}\int_X f(x)w(x)\,d\mu(x)\right) \leq \int_Z \varphi\left(\frac{1}{W}\int_X f(x)\alpha(x,z)w(x)\,d\mu(x)\right)d\lambda(z)$$
$$\leq \frac{1}{W}\int_X (\varphi \circ f)(x)w(x)\,d\mu(x), \qquad (2.15)$$

where $W = \int_X w(x)\,d\mu(x) > 0$.
If φ is concave, then the reversed signs of the inequalities hold in (2.15).

If λ is a discrete measure, then the refinement of the Jensen inequality was rediscovered recently, see, e.g., [54], while similar results can be found in [36, 49, 108] for some particular cases of α.

The simplest form of the reverse Jensen inequality is the following inequality where one weight is positive, while the second one is negative:

Let φ be a real convex function on I. If p and q are positive numbers such that $p - q > 0$, then

$$(p-q)\varphi\left(\frac{pa-qb}{p-q}\right) \geq p\varphi(a) - q\varphi(b) \qquad (2.16)$$

for all $a, b \in I$ such that $\frac{pa-qb}{p-q} \in I$.

This follows from the definition of a convex function: $\varphi(tx+(1-t)y) \leq t\varphi(x)+(1-t)\varphi(y)$, $t \in [0,1]$, $x, y \in I$ after the substitutions:

$$t = \frac{p-q}{p}, \quad x = \frac{pa-qb}{p-q}, \quad y = b.$$

The reverse Jensen inequality for integrals follows from Lemma 4.25 in the book [110, p. 124] and has the following form:

Theorem 2.9 *Let (X, μ) be a measure space. Let w be a weight and $w_0, f_0 \in \mathbb{R}$, $w_0 > W = \int_X w(x) d\mu(x) > 0$. Let φ be a real convex function on an interval I and $f_0 \in I$. Let f be a function on X such that wf and $w(\varphi \circ f)$ are integrable and $\frac{w_0 f_0 - \int_X f w \, d\mu}{w_0 - W} \in I$. Then*

$$(w_0 - W) \cdot \varphi\left(\frac{w_0 f_0 - \int_X w(x) f(x) \, d\mu(x)}{w_0 - W}\right)$$

$$\geq w_0 \varphi(f_0) - \int_X w(x)(\varphi \circ f)(x) \, d\mu(x). \qquad (2.17)$$

If φ is concave, then the reversed inequality holds.

Here we give a proof of Theorem 2.9 since we will use one step of that proof in our further investigation. We follow the proof in [110].

Proof By putting in (2.16)

$$p = w_0, \quad q = W, \quad a = f_0, \quad b = \frac{1}{W} \int_X f(x) w(x) \, d\mu(x),$$

we obtain that

$$(w_0 - W) \cdot \varphi\left(\frac{w_0 f_0 - \int_X w(x) f(x) \, d\mu(x)}{w_0 - W}\right)$$

$$\geq w_0 \varphi(f_0) - W \varphi\left(\frac{1}{W} \int_X f(x) w(x) \, d\mu(x)\right)$$

$$\geq w_0 \varphi(f_0) - \int_X w(x)(\varphi \circ f)(x) \, d\mu(x), \qquad (2.18)$$

where in the last inequality we use the Jensen inequality for integrals. The proof is complete. □

2.3 Refinements of the Jensen Inequality and Its Reversed Version

The following theorem is a continuous refinement of the previously mentioned reverse Jensen inequality (2.17) for integrals (see [102]).

Theorem 2.10 *Let the assumptions of Theorem* 2.8 *hold. Additionally, let $w_0 \in \mathbb{R}$ be such that $w_0 > W > 0$. Let φ be a real convex function on an interval I and $f_0 \in I$. Let f be a function on X such that wf and $w(\varphi \circ f)$ are integrable and $\frac{w_0 f_0 - \int_X w(x)f(x)\,d\mu(x)}{w_0 - W} \in I$. Then*

$$(w_0 - W) \cdot \varphi\left(\frac{w_0 f_0 - \int_X f(x) w(x) \, d\mu(x)}{w_0 - W}\right)$$

$$\geq w_0 \varphi(f_0) - W \int_Z \varphi\left(\frac{1}{W} \int_X f(x)\alpha(x,z) w(x) \, d\mu(x)\right) d\lambda(z)$$

$$\geq w_0 \varphi(f_0) - \int_X w(x)(\varphi \circ f)(x) \, d\mu(x). \qquad (2.19)$$

If φ is concave, then the reversed signs of the inequalities in (2.19) *hold.*

Proof Using the first inequality in (2.18) and (2.15), we get

$$(w_0 - W) \cdot \varphi\left(\frac{w_0 f_0 - \int_X f(x) w(x) \, d\mu(x)}{w_0 - W}\right)$$

$$\geq w_0 \varphi(f_0) - W \varphi\left(\frac{1}{W} \int_X f(x) w(x) \, d\mu(x)\right)$$

$$\geq w_0 \varphi(f_0) - W \int_Z \varphi\left(\frac{1}{W} \int_X f(x)\alpha(x,z) w(x) \, d\mu(x)\right) d\lambda(z)$$

$$\geq w_0 \varphi(f_0) - \int_X w(x)(\varphi \circ f)(x) \, d\mu(x)$$

and the proof is complete. □

It is known that the integral Hölder inequality for two functions is a consequence of the Jensen inequality for an appropriate function φ. It is to be expected that some already obtained results will also appear as consequences of the Jensen inequality. For a more consequent development of this idea, see [85], [110], and [113].

If $r, s > 1$ are numbers such that $\frac{1}{r} + \frac{1}{s} = 1$, then (2.15) for the concave function $\varphi(x) = x^{1/r}$ with the substitutions $w = wg^s$, $f = f^r g^{-s}$, where α satisfies assumptions (2.13) and (2.14), and $\int_X wg^s \alpha \, d\mu = \int_X wg^s \, d\mu$, gives the following continuous refinement of the Hölder inequality:

$$\|fg\|_{1,w} \leq \int_Z \|\alpha^{1/r}(\cdot,z)f\|_{r,w} \cdot \|\alpha^{1/s}(\cdot,z)g\|_{s,w} d\lambda$$
$$\leq \|f\|_{r,w} \cdot \|g\|_{s,w}. \tag{2.20}$$

As usual, by $\|F\|_{p,w}$ we denote: $\|F\|_{p,w} = \left(\int_X |F(x)|^p w(x)\, d\mu(x)\right)^{1/p}$. The particular case of (2.20) when the continuous refinement collapses to the sum of two functions u, v, such that $u(x) + v(x) = 1$ on $X = [a,b]$, was derived in [54].

Example 2.2 Putting in (2.19) substitutions: $\varphi(x) = x^{1/r}$, $w = wg^s$, $f = f^r g^{-s}$, $w_0 = w_0 c_2^s$, $f_0 = c_1^r c_2^{-s}$, and if α satisfies assumptions (2.13) and (2.14), and $\int_X wg^s \alpha\, d\mu = \int_X wg^s\, d\mu$, then we have the following refinement of the Popoviciu inequality:

$$w_0 c_1 c_2 - \|fg\|_{1,w} \geq w_0 c_1 c_2 - \int_Z \|\alpha^{1/r}(\cdot,z)f\|_{r,w} \|\alpha^{1/s}(\cdot,z)g\|_{s,w} d\lambda$$
$$\geq \left(w_0 c_1^r - \|f\|_{r,w}^r\right)^{1/r} \left(w_0 c_2^s - \|g\|_{s,w}^s\right)^{1/s},$$

provided that all integrals exist.

Finally, we also point out that by using the continuous refinement (2.15) of the Jensen inequality, we can derive some different refinements of the Hölder and the Popoviciu inequalities.

Theorem 2.11 *Let $r,s > 1$ be numbers such that $\dfrac{1}{r} + \dfrac{1}{s} = 1$. Let (X,μ) and (Z,λ) be two measure spaces, $\int_Z d\lambda = 1$, $w : X \to [0,\infty)$ be a measurable mapping on X such that $\int_X w\, d\mu \neq 0$, and $\alpha : X \times Z \to [0,\infty)$ be a function which satisfies (2.14). Let $c_1, c_2, w_0 > 0$ and $f, g : X \to [0,\infty)$ be such that $w_0 c_1^r - \|f\|_{r,w}^r > 0$, $w_0 c_2^s - \|g\|_{s,w}^s > 0$, and $\int_X \alpha(x,z) w(x) g^s(x)\, d\mu(x) = \int_X w(x) g^s(x)\, d\mu(x)$, $z \in Z$, hold. Then:*

(i) The following continuous refinement of the Hölder inequality holds:

$$\|fg\|_{1,w} \leq \left(\int_Z \|fg\alpha\|_{1,w}^r d\lambda\right)^{\frac{1}{r}} \leq \|f\|_{r,w} \cdot \|g\|_{s,w}, \tag{2.21}$$

provided that all integrals exist.

(ii) The following continuous refinement of the Popoviciu inequality holds:

$$w_0 c_1 c_2 - \|fg\|_{1,w}$$
$$\geq \left(w_0 c_2^s - \|g\|_{s,w}^s\right)^{\frac{1}{s}} \left(w_0 c_1^r - \frac{1}{\|g\|_{s,w}^r} \int_Z \|fg\alpha(\cdot, z)\|_{1,w}^r \, d\lambda(z)\right)^{\frac{1}{r}}$$
$$\geq \left(w_0 c_1^r - \|f\|_{r,w}^r\right)^{\frac{1}{r}} \left(w_0 c_2^s - \|g\|_{s,w}^s\right)^{\frac{1}{s}}, \qquad (2.22)$$

provided that all integrals exist.

Proof

(i) By making the substitutions

$$\varphi(x) = x^r, \quad w = wg^s, \quad \text{and} \quad f = fg^{-s/r}$$

in (2.15) for the convex function φ, we get inequality (2.21).

(ii) After the substitutions

$$w_0 = w_0 c_2^s, \quad f_0 = c_1 c_2^{-s/r} \quad w = wg^s, \quad \text{and} \quad f = fg^{-s/r}$$

in (2.19) for the convex function $\varphi(x) = x^r$, we get the inequality (2.22). The proof is complete. □

2.4 Refinement of the Jensen-Mercer Inequality

In [78] the following variant of the Jensen inequality was proved.

Theorem 2.12 *Let $[a, b] \subset \mathbb{R}$ and $x_1, \ldots, x_n$ be from $[a, b]$. Suppose that $(p_1, \ldots, p_n)$ is a nonnegative real n-tuple such that $P_n = \sum_{i=1}^n p_i > 0$. If $\varphi : [a, b] \to \mathbb{R}$ is a convex function, then the following inequality holds:*

$$\varphi\left(a + b - \frac{1}{P_n} \sum_{i=1}^n p_i x_i\right) \leq \varphi(a) + \varphi(b) - \frac{1}{P_n} \sum_{i=1}^n p_i \varphi(x_i). \qquad (2.23)$$

Nowadays, the above inequality is known as the discrete Jensen-Mercer inequality. Further studying of this result leads to its integral version, which was given in [32], and we state it also here.

Theorem 2.13 *Let $[a, b] \subset \mathbb{R}$ and let $\varphi : [a, b] \to \mathbb{R}$ be a continuous convex function. Let (X, μ) be a probability measure space and $f : X \to [a, b]$ be an integrable function on X. Then*

$$\varphi\left(a + b - \int_X f(x)\, d\mu(x)\right) \leq \varphi(a) + \varphi(b) - \int_X \varphi(f(x))\, d\mu(x). \quad (2.24)$$

It is easy to see that for a (non-probability) measure space (X, μ) with $0 < P = \int_X p(x) d\mu(x) < \infty$, where p is a nonnegative integrable function on X, (2.24) implies the following weighted version of the Jensen-Mercer inequality for integrals:

$$\varphi\left(a + b - \frac{1}{P}\int_X p(x)f(x)\, d\mu(x)\right)$$
$$\leq \varphi(a) + \varphi(b) - \frac{1}{P}\int_X p(x)\varphi(f(x))\, d\mu(x). \quad (2.25)$$

Here we give a continuous refinement of inequality (2.25).

Theorem 2.14 *Let (X, μ) and (Z, λ) be two measure spaces, $\int_Z d\lambda(z) = 1$, and $p : X \to [0, \infty)$ be an integrable mapping on X such that $P = \int_X p(x)\, d\mu(x) \neq 0$, let $\alpha : X \times Z \to [0, \infty)$ be a measurable function on $X \times Z$ satisfying*

$$\frac{1}{P}\int_X \alpha(x, z)p(x)\, d\mu(x) = 1, \quad \text{for each } z \in Z \quad (2.26)$$

and

$$\int_Z \alpha(x, z)\, d\lambda(z) = 1, \quad \text{for each } x \in X. \quad (2.27)$$

If φ is a continuous convex function on the interval $[a, b] \subset \mathbb{R}$, then for the integrable function $f : X \to [a, b]$, it holds that

$$\varphi\left(a + b - \frac{1}{P}\int_X p(x)f(x)\, d\mu(x)\right)$$
$$\leq \int_Z \varphi\left(a + b - \frac{1}{P}\int_X p(x)\alpha(x, z)f(x)\, d\mu(x)\right) d\lambda(z)$$
$$\leq \varphi(a) + \varphi(b) - \frac{1}{P}\int_X p(x)\varphi(f(x))\, d\mu(x). \quad (2.28)$$

Proof As we know, the continuous refinement of Jensen inequality involving a function $\bar{f}$ states (see Theorem 2.8):

2.4 Refinement of the Jensen-Mercer Inequality

$$\varphi\left(\frac{1}{P}\int_X p(x)\bar{f}(x)\,d\mu(x)\right) \leq \int_Z \varphi\left(\frac{1}{P}\int_X p(x)\alpha(x,z)\bar{f}(x)\,d\mu(x)\right)d\lambda(z)$$
$$\leq \frac{1}{P}\int_X p(x)\varphi(\bar{f}(x))\,d\mu(x), \quad (2.29)$$

where (2.26) and (2.27) are satisfied. For $\bar{f} = a + b - f$ we obtain

$$\varphi\left(\frac{1}{P}\int_X p(x)(a+b-f(x))\,d\mu(x)\right)$$
$$\leq \int_Z \varphi\left(\frac{1}{P}\int_X p(x)\alpha(x,z)(a+b-f(x))\,d\mu(x)\right)d\lambda(z)$$
$$\leq \frac{1}{P}\int_X p(x)\varphi(a+b-f(x))\,d\mu(x). \quad (2.30)$$

Using the assumptions, we get

$$\varphi\left(\frac{1}{P}\int_X p(x)(a+b-f(x))\,d\mu(x)\right) = \varphi\left(a+b-\frac{1}{P}\int_X p(x)f(x)\,d\mu(x)\right)$$

and

$$\frac{1}{P}\int_X p(x)\alpha(x,z)(a+b-f(x))\,d\mu(x)$$
$$= a+b-\frac{1}{P}\int_X p(x)\alpha(x,z)f(x)\,d\mu(x). \quad (2.31)$$

Moreover, we note that inequality (2.23) for $n = 1$ and $p_1 = 1$ collapses to $\varphi(a+b-f(x)) \leq \varphi(a) + \varphi(b) - \varphi(f(x))$. Hence,

$$\frac{1}{P}\int_X p(x)\varphi(a+b-f(x))\,d\mu(x)$$
$$\leq \varphi(a) + \varphi(b) - \frac{1}{P}\int_X p(x)\varphi(f(x))\,d\mu(x). \quad (2.32)$$

Finally, by combining (2.29)–(2.32), we can conclude that (2.28) holds, so the proof is complete. □

2.5 Refinements of the Hardy Inequality

The classical Hardy inequality reads

$$\int_0^\infty \left(\frac{1}{x}\int_0^x f(y)\,dy\right)^p dx \le \left(\frac{p}{p-1}\right)^p \int_0^\infty f^p(y)\,dy, \quad p > 1, \qquad (2.33)$$

where f is a nonnegative function, $f \in L_p(0, \infty)$. This inequality was proved by G. H. Hardy in his famous paper [46] after almost 10 years of research. This dramatic prehistory was described in the paper [66]. After that, it has been almost 100 years of research in this fascinating area. This development up to 2007 was described in the monograph [67] and the further development up to 2017 in the monograph [69]. See also the monographs [56], [64], [105] and the references in all these monographs. One important such development is when the classical Hardy operator $Hf(x) = \frac{1}{x}\int_0^x f(y)\,dy$ is replaced by the kernel Hardy operator H_k where

$$H_k f(x) = \frac{1}{K(x)} \int_0^x k(x, y) f(y)\,dy, \qquad (2.34)$$

with $K(x) = \int_0^x k(x, y)\,dy$.

In particular, Chapter 2 and Section 7.5 in [69] are devoted to this subject. Here we just note that such a development may be regarded as continuous Hardy-type inequalities. In order to illustrate this idea we first present the following result which was stated and proved in [62] (see also [61, Theorem 2.5]).

Theorem 2.15 *Assume that* (X, μ) *and* (Y, ν) *are two measure spaces. Let* $k : X \times Y \to \mathbb{R}$ *be a nonnegative measurable function on* $X \times Y$ *with*

$$K(x) = \int_Y k(x, y)\,d\nu(y) < \infty, \quad x \in X.$$

Let $u : X \to [0, \infty)$ *be integrable and denote*

$$\beta(y) = \int_X \frac{k(x, y)u(x)}{K(x)}\,d\mu(x), \quad y \in Y.$$

If $\phi : I \to \mathbb{R}$ *is convex, then*

$$\int_X u(x)\phi\left(\int_Y \frac{k(x, y)}{K(x)} f(y)\,d\nu(y)\right) d\mu(x) \le \int_Y \beta(y)\phi(f(y))\,d\nu(y) \qquad (2.35)$$

holds for all measurable functions f *on* Y, *such that* $\operatorname{Im} f \subseteq I$.

2.5 Refinements of the Hardy Inequality

Example 2.3 For $X = Y = [0, \infty)$, $u(x) = \frac{1}{x}$, $k(x, y) = 1$ for $y \leq x$, and 0 otherwise, inequality (2.35) becomes

$$\int_0^\infty \phi\left(\frac{1}{x}\int_0^x f(y)\,dy\right)\frac{dx}{x} \leq \int_0^\infty \phi(f(y))\frac{dy}{y}, \tag{2.36}$$

which collapses to the classical Hardy inequality (2.33) after substitutions $\phi(x) = x^p$, $p > 1$, and $g(x) = f(x^{\frac{p-1}{p}})x^{-\frac{1}{p}}$.

Remark 2.5 Nowadays often (2.36) is called a fundamental form of the Hardy inequality since it shows its convexity character. For example, by instead doing the substitution $\phi(x) = x^p$, $p \geq 1$ or $p < 0$, $f(x) = g\left(x^{\frac{p-1-a}{p}}\right)x^{-\frac{1+a}{p}}$, $a < p - 1$, we find that (2.36) is also equivalent to that

$$\int_0^\infty \left(\frac{1}{x}\int_0^x g(y)\,dy\right)^p x^a\,dx \leq \left(\frac{p}{p-1-a}\right)^p \int_0^\infty g^p(x)x^a\,dx, \tag{2.37}$$

i.e., the first proved weighted Hardy inequality holds (and even for $p < 0$). Moreover, (2.36) holds in the reversed direction for ϕ concave, which in the same way implies the reversed Hardy inequality (2.37) holds for $0 < p < 1$, $a > p - 1$, with the sharp constant $\left(\frac{p}{a+1-p}\right)^p$. This fact follows since the Jensen inequality holds in the reversed direction for the concave function $\phi(x) = x^p$, $0 < p < 1$, and by doing similar substitutions as above.

Next we present a continuous refinement of inequality (2.35).

Theorem 2.16 *Let assumptions of Theorem* 2.15 *hold. Let* (Z, λ) *be a measure space where* $\int_Z d\lambda(z) = 1$. *Let* $\alpha : Y \times Z \to [0, \infty)$ *be a function such that*

$$\int_Z \alpha(y, z)\,d\lambda(z) = 1, \ y \in Y, \quad \int_Y \alpha(y, z)\frac{k(x, y)}{K(x)}\,d\nu(y) = 1, \ z \in Z.$$

Then, for the convex function ϕ we have

$$\int_X u(x)\phi\left(\frac{1}{K(x)}\int_Y k(x, y)f(y)\,d\nu(y)\right)d\mu(x)$$
$$\leq \int_X u(x)\left(\int_Z \phi\left(\frac{1}{K(x)}\int_Y k(x, y)\alpha(y, z)f(y)\,d\nu(y)\right)d\lambda(z)\right)d\mu(x)$$
$$\leq \int_Y \beta(y)\phi(f(y))\,d\nu(y), \tag{2.38}$$

where

$$\beta(y) = \int_X u(x) \frac{k(x,y)}{K(x)} \, d\mu(x), \quad y \in Y. \qquad (2.39)$$

Moreover, if ϕ is concave, then both inequalities hold in the reversed direction.

Proof We denote $v(x,y) = \frac{k(x,y)}{K(x)}$. Putting in the continuous refinement of the Jensen inequality (see Theorem 2.8):

$$X = Y, \quad w(x) = v(x,y), \quad \text{and} \quad \mu(x) = \nu(y),$$

we get for fixed x:

$$\phi\left(\int_Y f(y)v(x,y)\,d\nu(y)\right) \leq \int_Z \phi\left(\int_Y f(y)\alpha(y,z)v(x,y)\,d\nu(y)\right) d\lambda(z)$$
$$\leq \int_Y \phi(f(y))v(x,y)\,d\nu(y). \qquad (2.40)$$

Multiplying inequality (2.40) by $u(x)$, integrating over X, and applying the Fubini theorem, we obtain

$$\int_X u(x)\phi\left(\int_Y f(y)v(x,y)\,d\nu(y)\right) d\mu(x)$$
$$\leq \int_X u(x)\left(\int_Z \phi\left(\int_Y f(y)\alpha(y,z)v(x,y)\,d\nu(y)\right) d\lambda(z)\right) d\mu(x)$$
$$\leq \int_X u(x)\left(\int_Y \phi(f(y))v(x,y)\,d\nu(y)\right) d\mu(x)$$
$$= \int_Y \phi(f(y))\left(\int_X u(x)v(x,y)\,d\mu(x)\right) d\nu(y) = \int_Y \phi(f(y))\beta(y)\,d\nu(y).$$

Hence, (2.38) holds. The proof of the case when ϕ is concave is completely similar, so we omit the details. The proof is complete. □

By applying Theorem 2.16 with $\phi(t) = t^p$, we get the following result:

Corollary 2.2 *Let the assumptions of Theorem 2.16 hold. Then, for $p \geq 1$ or $p < 0$ the following refinement (of the Hardy-type inequality (2.42) below) holds:*

$$\int_X u(x)\left(\frac{1}{K(x)}\int_Y k(x,y)f(y)\,d\nu(y)\right)^p d\mu(x)$$

2.5 Refinements of the Hardy Inequality

$$\leq \int_X u(x) \left(\int_Z \left(\frac{1}{K(x)} \int_Y k(x,y) \alpha(y,z) f(y) \, d\nu(y) \right)^p d\lambda(z) \right) d\mu(x)$$

$$\leq \int_Y f^p(y) \beta(y) \, d\nu(y), \tag{2.41}$$

where β is defined by (2.39) and provided that the involved integrals exist. For the case $p < 0$ we must also have the obvious restriction that the expressions within the brackets are strictly greater than 0.

Moreover, if $0 < p < 1$, then both inequalities hold in the reversed direction (if $p = 1$ we have even equality).

In particular, in case when $p \geq 1$ or $p < 0$, (2.41) gives a refinement of the following Hardy-type inequality (and its reversed version for $0 < p < 1$):

$$\int_X u(x) \left(\frac{1}{K(x)} \int_Y k(x,y) f(y) \, d\nu(y) \right)^p d\mu(x) \leq \int_Y f^p(y) \beta(y) \, d\nu(y) \tag{2.42}$$

under the restrictions given in Theorem 2.15.

In the next theorem we state another refinement of inequality (2.42).

Theorem 2.17 *Assume that (X, μ) and (Y, ν) are two measure spaces. Let f be a nonnegative function on Y, $f \in L_p(Y)$ ($p > 1$), and $k : X \times Y \to \mathbb{R}$ be a nonnegative measurable function on $X \times Y$ with*

$$K(x) = \int_Y k(x,y) \, d\nu(y) < \infty, \quad x \in X.$$

Let $u : X \to [0, \infty)$ be integrable and β is defined by (2.39). Let (Z, dz) be a measure space, and $\alpha(y,z)$ be a nonnegative integrable function on $Y \times Z$ such that $\int_Z \alpha(z,y) \, dz = 1$ for any $y \in Y$.

Then for $p > 1$ the following refinement of the Hardy-type inequality (2.42) holds:

$$\int_X u(x) \left(\frac{1}{K(x)} \int_Y f(y) k(x,y) \, d\nu(y) \right)^p d\mu(x)$$

$$\leq \int_X u(x) \left\{ \frac{1}{K(x)} \int_Z \left(\int_Y \alpha(y,z) f^p(y) k(x,y) \, d\nu(y) \right)^{1/p} \times \right.$$

$$\left. \times \left(\int_Y \alpha(y,z) k(x,y) \, d\nu(y) \right)^{\frac{p-1}{p}} dz \right\}^p d\mu(x)$$

$$\leq \int_Y f^p(y) \beta(y) \, d\nu(y). \tag{2.43}$$

Proof We use abbreviations f for $f(y)$, etc. Moreover, as before we put $v(x, y) = \frac{k(x,y)}{K(x)}$. Let us recall the refinement of the Hölder inequality given in (2.5):

$$\int_Y fgv\,dv \leq \int_Z \left(\int_Y \alpha(y, z) f^p v\,dv\right)^{1/p} \left(\int_Y \alpha(y, z) g^q v\,dv\right)^{1/q} dz$$

$$\leq \left(\int_Y f^p v\,dv\right)^{1/p} \left(\int_Y g^q v\,dv\right)^{1/q}, \tag{2.44}$$

where $p, q > 1$ such that $\frac{1}{p} + \frac{1}{q} = 1$. Putting in (2.44): $g(y) = 1$ and using the fact that $\int_Y v\,dv = 1$, we have:

$$\int_Y fv\,dv \leq \int_Z \left(\int_Y \alpha f^p v\,dv\right)^{1/p} \left(\int_Y \alpha v\,dv\right)^{1/q} dz$$

$$\leq \left(\int_Y f^p v\,dv\right)^{1/p} \left(\int_Y v\,dv\right)^{1/q} = \left(\int_Y f^p v\,dv\right)^{1/p}.$$

Now take the pth power of the previous inequality, multiply with u and integrate over X, use the Fubini theorem, and we find that

$$\int_X u \left(\int_Y fv\,dv\right)^p d\mu$$

$$\leq \int_X u \left\{\int_Z \left(\int_Y \alpha f^p v\,dv\right)^{1/p} \left(\int_Y \alpha v\,dv\right)^{1/q} dz\right\}^p d\mu$$

$$\leq \int_X u \left\{\left(\int_Y f^p v\,dv\right)^{1/p}\right\}^p d\mu = \int_X u \left(\int_Y f^p v\,dv\right) d\mu$$

$$= \int_Y f^p \left(\int_X u \frac{k(x, y)}{K(x)} d\mu\right) dv = \int_Y f^p \beta\,dv,$$

which means that (2.43) holds, so the proof is complete. □

Remark 2.6 For more complementary refinements of continuous forms of Hardy-type inequalities, we refer to the interesting PhD thesis [61], see also our Sect. 3.5.

Remark 2.7 An interesting question in the theory of Hardy-type inequalities is the following (still open) question: Characterize

$$\|Tf\|_{q,u} \leq C\|f\|_{p,v}, \tag{2.45}$$

2.5 Refinements of the Hardy Inequality

where u and v are weight functions and $p, q \geq 1$. Here

$$Tf(x) = \int_a^x k(x, y) f(y) \, dy,$$

where $k(x, y)$ denotes a general positive kernel:

(i) Without restrictions on the kernel $k(x, y)$ this long-lasting problem is still open.
(ii) The solution of this problem is known for a number of special cases and parameters.

We just mention the following result which can be a step toward solving this open question in the convexity case $1 < p \leq q < \infty$ (see [68]):

Theorem 2.18 *Let $1 < p \leq q < \infty$, $a < b \leq \infty$, u, v be weights, and let $k(x, y)$ be a nonnegative kernel.*

(a) Then (2.45) holds if

$$A_s = \sup_{a < y < b} \left(\int_y^b k^q(x, y) u(x) V^{\frac{q(p-s-1)}{p}}(x) \, dx \right)^{1/q} V^{s/p}(y) < \infty, \quad (2.46)$$

for any $s < p - 1$.
(b) The condition (2.46) cannot be improved in general for $s > 0$ because for product kernels it is even necessary and sufficient for (2.45) to hold.
(c) For the best constant C in (2.45) we have the following estimate:

$$C \leq \inf_{s < p-1} \left(\frac{p}{p-s-1} \right)^{1/p'} A_s.$$

Here, as usual,

$$V(x) = \int_a^x v^{1-p'}(y) \, dy \quad \text{and} \quad \frac{1}{p} + \frac{1}{p'} = 1.$$

Remark 2.8 This result opens a possibility that the condition (2.46) can be candidate to solve the open question we have pointed out in Remark 2.7(i) above.

3. Refinements of Inequalities via Strong Convexity and Superquadracity

This chapter also contains refinements of some continuous forms of classical inequalities, but now we pay attention to results where, instead of convex functions, we use the following variants: the class of strongly convex functions and the class of superquadratic functions. The results in this chapter are mainly published in the papers [102] and [103].

3.1 On the Jensen Inequality for Strongly Convex Functions

One of our objectives is to give results for strongly convex functions. First we define and present some useful facts about that class of functions.

Definition 3.1 Let I be an interval of the real line. A function $\varphi : I \to \mathbb{R}$ is called a strongly convex function with modulus $c > 0$ if

$$\varphi(tu + (1-t)v) \leq t\varphi(u) + (1-t)\varphi(v) - ct(1-t)(u-v)^2$$

for all $u, v \in I$ and all $t \in [0, 1]$.

The theory of strongly convex functions is vast, but here we point out only a very useful characterization of it. Namely, a function φ is strongly convex with modulus $c > 0$ if and only if the function $\psi(x) = \varphi(x) - cx^2$ is convex (see [85, 86]). This fact plays a crucial role in proofs of our results.

The Jensen-type inequality for strongly convex functions was given in its discrete and integral forms in [79], and the Jensen-type inequality for integrals reads:

Theorem 3.1 *Let (X, μ) be a probability measure space, and I be an open interval in $\mathbb{R}$. Let $\varphi : I \to \mathbb{R}$ be a strongly convex function with modulus $c > 0$ and let $f : X \to I$ be an integrable function, $f^2 \in L_1(X)$. Then*

$$\varphi\left(\int_X f \, d\mu\right) \leq \int_X (\varphi \circ f) \, d\mu - c \int_X (f - \bar{f})^2 \, d\mu, \qquad (3.1)$$

where $\bar{f} = \int_X f \, d\mu$.

The following results, published in [102], give us refinements of the continuous Jensen inequality for strongly convex functions and its reverse. Moreover, refinements of the Hermite-Hadamard and Lah-Ribarič inequality are given.

Theorem 3.2 *Let (X, μ) and (Z, λ) be two probability measure spaces, and $\alpha : X \times Z \to [0, \infty)$ be a measurable function satisfying*

$$\int_X \alpha(x, z) \, d\mu(x) = 1 \quad \text{for each} \quad z \in Z \qquad (3.2)$$

and

$$\int_Z \alpha(x, z) \, d\lambda(z) = 1 \quad \text{for each} \quad x \in X. \qquad (3.3)$$

If $\varphi : I \to \mathbb{R}$ is a strongly convex function with modulus $c > 0$ and $f : X \to I$ is an integrable function, $f^2 \in L_1(X)$, then

$$\varphi\left(\int_X f \, d\mu\right) \leq \int_Z \varphi\left(\int_X f(x)\alpha(x, z) \, d\mu(x)\right) d\lambda(z)$$
$$- c \int_Z \left(\int_X f(x)\alpha(x, z) \, d\mu(x) - \bar{f}\right)^2 d\lambda(z)$$
$$\leq \int_X (\varphi \circ f) \, d\mu - c \int_X (f - \bar{f})^2 \, d\mu, \qquad (3.4)$$

where $\bar{f} = \int_X f \, d\mu$.

Proof Since the function φ is strongly convex, the function $\varphi(x) - cx^2$ is convex, and for this function, the refinement (2.15) holds. Therefore, after adding the term $c\left(\int_X f \, d\mu\right)^2 = c\bar{f}^2$ on each side of (2.15), we get

3.1 On the Jensen Inequality for Strongly Convex Functions

$$\varphi\left(\int_X f \, d\mu\right) \leq \int_Z \varphi\left(\int_X f(x)\alpha(x, z) \, d\mu(x)\right) d\lambda(z)$$
$$- c\left[\int_Z \left(\int_X f(x)\alpha(x, z) \, d\mu(x)\right)^2 d\lambda(z) - \bar{f}^2\right]$$
$$\leq \int_X (\varphi \circ f) \, d\mu - c\left[\int_X f^2 \, d\mu - \bar{f}^2\right]. \tag{3.5}$$

We use the assumptions on μ and transform the term $\int_X f^2 \, d\mu - \bar{f}^2$ as follows:

$$\int_X f^2 \, d\mu - \bar{f}^2 = \int_X f^2 \, d\mu - 2\bar{f} \cdot \bar{f} + \bar{f}^2$$
$$= \int_X f^2 \, d\mu - 2\int_X \bar{f} f \, d\mu + \int_X \bar{f}^2 \, d\mu = \int_X (f - \bar{f})^2 \, d\mu.$$

We denote $F(z) = \int_X f(x)\alpha(x, z) \, d\mu(x)$. Using the Fubini theorem and the properties of the weight α, we obtain

$$\bar{F} = \int_Z F(z) \, d\lambda(z) = \int_Z \left(\int_X f(x)\alpha(x, z) \, d\mu(x)\right) d\lambda(z)$$
$$= \int_X \left(\int_Z \alpha(x, z) \, d\lambda(z)\right) f(x) d\mu(x) = \int_X 1 \cdot f(x) d\mu(x) = \bar{f}. \tag{3.6}$$

With this notation and using the same method as above, we find that the second term in the middle expression of (3.5) is equal to

$$c\left[\int_Z F^2(z) d\lambda(z) - \bar{F}^2\right] = c \int_Z \left[\int_X f(x)\alpha(x, z) d\mu(x) - \bar{f}\right]^2 d\lambda(z). \tag{3.7}$$

By combining (3.5)–(3.7), we obtain (3.4) and the proof is complete. □

Remark 3.1 Since $c > 0$ the chain of inequalities in (3.4) can be followed by $\leq \int_X (\varphi \circ f) \, d\mu$. Hence, Theorem 3.2 is indeed a genuine refinement of the Jensen inequality.

It is interesting to state the corresponding refinements for some particular cases such as for discrete and for integral Jensen's inequality with finitely many functions involved.

Corollary 3.1 *(i) Let $-\infty \leq a < b \leq \infty$, $\varphi : I \to \mathbb{R}$ be a strongly convex function with modulus c. Let $w, f, \alpha_i : [a, b] \to \mathbb{R}$, $i = 1, 2, \ldots, n$, be integrable functions such* that $w, \alpha_i \geq 0$, $\sum_{i=1}^{n} \alpha_i(x) = 1$ for each $x \in [a, b]$, $\int_a^b w \, dx \neq 0$, $\int_a^b \alpha_i w \, dx \neq 0$, $f^2 \in L_1(a, b)$, and $f([a, b]) \subseteq I$. Then the following refinement of the Jensen inequality holds:

$$\varphi\left(\frac{1}{W}\int_a^b fw \, dx\right) \leq \frac{1}{W}\sum_{i=1}^{n}\left(\int_a^b \alpha_i w \, dx\right) \varphi\left(\frac{\int_a^b \alpha_i fw \, dx}{\int_a^b \alpha_i w \, dx}\right)$$

$$- \frac{c}{W}\sum_{i=1}^{n}\left(\int_a^b \alpha_i w \, dx\right)\left(\frac{\int_a^b \alpha_i fw \, dx}{\int_a^b \alpha_i w \, dx} - \bar{f}\right)^2$$

$$\leq \frac{1}{W}\int_a^b (\varphi \circ f)w \, dx - \frac{c}{W}\int_a^b (f - \bar{f})^2 w \, dx$$

$$\leq \frac{1}{W}\int_a^b (\varphi \circ f)w \, dx, \qquad (3.8)$$

where $W = \int_a^b w \, dx$ and $\bar{f} = \frac{1}{W}\int_a^b fw \, dx$.

(ii) Let w_j, $j = 1, \ldots, m$ be nonnegative numbers such that $\sum_{j=1}^{m} w_j \neq 0$, let α_{ij}, $i = 1, \ldots, n$, $j = 1, \ldots, m$ be nonnegative numbers such that $\sum_{j=1}^{m} w_j \alpha_{ij} \neq 0$, $i = 1, \ldots, n$, and $\sum_{i=1}^{n} \alpha_{ij} = 1$, $j = 1, \ldots, m$. Let f_j, $j = 1, \ldots, m$, be real numbers from an interval I. Then, for any strongly convex function $\varphi : I \to \mathbb{R}$ with modulus c the following refinement of the discrete Jensen inequality holds:

$$\varphi\left(\frac{1}{W}\sum_{j=1}^{m} w_j f_j\right) \leq \frac{1}{W}\sum_{i=1}^{n}\left(\sum_{j=1}^{m} w_j \alpha_{ij}\right) \varphi\left(\frac{\sum_{j=1}^{m} w_j \alpha_{ij} f_j}{\sum_{j=1}^{m} w_j \alpha_{ij}}\right)$$

$$- \frac{c}{W}\sum_{i=1}^{n}\left(\sum_{j=1}^{m} w_j \alpha_{ij}\right)\left(\frac{\sum_{j=1}^{m} w_j \alpha_{ij} f_j}{\sum_{j=1}^{m} w_j \alpha_{ij}} - \bar{f}\right)^2$$

$$\leq \frac{1}{W}\sum_{j=1}^{m} w_j \varphi(f_j) - \frac{c}{W}\sum_{j=1}^{m} w_j (f_j - \bar{f})^2$$

$$\leq \frac{1}{W}\sum_{j=1}^{m} w_j \varphi(f_j), \qquad (3.9)$$

where $W = \sum_{j=1}^{m} w_j$ and $\bar{f} = \frac{1}{W}\sum_{j=1}^{m} w_j f_j$.

3.1 On the Jensen Inequality for Strongly Convex Functions

Proof

(i) By applying Theorem 3.2 for

$$Z = Z_1 \cup \ldots \cup Z_n, \quad Z_i = [i-1, i), \quad Z_n = [n-1, n], \quad i = 1, 2, \ldots, n-1,$$

$$X = [a, b], \quad d\mu(x) = \frac{w(x)}{W}\, dx, \quad d\lambda(z) = \frac{1}{W}\left(\int_a^b \alpha_i(x) w(x)\, dx\right) dz \text{ for } z \in Z_i,$$

$$\alpha(x, z) = W \frac{\alpha_i(x)}{\int_a^b \alpha_i(x) w(x)\, dx} \quad \text{for } z \in Z_i,$$

then the inequalities in (3.4) imply those in (3.8).

(ii) By applying Theorem 3.2 for the same substitutions for Z and λ as we did in the proof of part (i), together with the following:

$$X = X_1 \cup \ldots \cup X_m, \quad X_i = [i-1, i), \quad X_m = [m-1, m], \quad i = 1, 2, \ldots, m-1,$$

$$d\mu(x) = \frac{w(x)}{W}\, dx, \quad w(x) = w_j, \quad f(x) = f_j, \quad \alpha_i(x) = \alpha_{ij} \quad \text{for } x \in X_j,$$

then the inequalities in (3.4) and a trivial argument give (3.9). The proof is complete. □

We also state the following refinement with finite partition of the space X:

Corollary 3.2 *Let the assumptions of Theorem 3.1 hold.*
Let $X_1, \ldots, X_n$ be a partition of the set X. Let μ have the additional property that $\int_{X_i} d\mu \neq 0$, $i = 1, 2, \ldots, n$.
Then, for any strongly convex function φ with modulus c the following refinement of the Jensen inequality holds:

$$\varphi\left(\int_X f\, d\mu\right) \leq \sum_{i=1}^n \left(\int_{X_i} d\mu\right) \varphi\left(\frac{\int_{X_i} f\, d\mu}{\int_{X_i} d\mu}\right)$$

$$- c \sum_{i=1}^n \left(\int_{X_i} d\mu\right) \left(\frac{\int_{X_i} f\, d\mu}{\int_{X_i} d\mu} - \bar{f}\right)^2$$

$$\leq \int_X (\varphi \circ f)\, d\mu - c \int_X (f - \bar{f})^2\, d\mu \leq \int_X (\varphi \circ f)\, d\mu, \qquad (3.10)$$

where $\bar{f} = \int_X f\, d\mu$.

Proof Let us use the same partition of the space Z as in the proof of Corollary 3.1, where the functions α_i are defined by

$$\alpha_i = \chi_{X_i}, \quad i = 1, 2, \ldots, n.$$

Here, χ_S denotes the characteristic function of the set S. Then the assumptions of Theorem 3.2 are satisfied. Hence, (3.4) and a trivial estimate show that the inequalities in (3.10) hold. The proof is complete. □

Example 3.1 If in Corollary 3.2 we put $X = [a, b]$, $a = a_0 < a_1 < a_2 < \ldots < a_n = b$, and $X_i = [a_{i-1}, a_i)$ for $i = 1, 2, \ldots, n$, $d\mu = \frac{w(x)}{W} dx$, then the inequalities in (3.10) become as follows:

$$\varphi\left(\frac{1}{W}\int_a^b fw\,dx\right) \leq \frac{1}{W}\sum_{i=1}^n \left(\int_{a_{i-1}}^{a_i} w\,dx\right) \varphi\left(\frac{\int_{a_{i-1}}^{a_i} fw\,dx}{\int_{a_{i-1}}^{a_i} w\,dx}\right)$$

$$- \frac{c}{W}\sum_{i=1}^n \left(\int_{a_{i-1}}^{a_i} w\,dx\right)\left(\frac{\int_{a_{i-1}}^{a_i} fw\,dx}{\int_{a_{i-1}}^{a_i} w\,dx} - \bar{f}\right)^2$$

$$\leq \frac{1}{W}\int_a^b (\varphi \circ f)w\,dx - \frac{c}{W}\int_a^b (f - \bar{f})^2 w\,dx$$

$$\leq \frac{1}{W}\int_a^b (\varphi \circ f)w\,dx, \qquad (3.11)$$

where $W = \int_a^b w\,dx$ and $\bar{f} = \frac{1}{W}\int_a^b fw\,dx$.

Finally, we state a refined version of the continuous reversed Jensen inequality (2.17) for strongly convex functions.

Theorem 3.3 *Let the assumptions of Theorems 2.8 and 2.10 hold. Then, for the strongly convex function φ with modulus c the following inequalities hold:*

$$(w_0 - W) \cdot \varphi\left(\frac{w_0 f_0 - \int_X fw\,d\mu}{w_0 - W}\right)$$

$$\geq w_0 \varphi(f_0) - W\int_Z \varphi\left(\frac{1}{W}\int_X f(x)\alpha(x,z)w(x)\,d\mu(x)\right) d\lambda(z)$$

$$- c\left(w_0 f_0^2 - W\int_Z \left(\frac{1}{W}\int_X f(x)\alpha(x,z)w(x)\,d\mu(x)\right)^2 d\lambda(z)\right)$$

$$-\frac{(w_0 f_0 - \int_X f w \, d\mu)^2}{w_0 - W}\Bigg)$$

$$\geq w_0 \varphi(f_0) - \int_X (\varphi \circ f) w \, d\mu - \frac{c}{w_0 - W} \int_X \left[W(f - \overline{f})^2 - w_0(f_0 - f)^2 \right] w \, d\mu,$$

where $\overline{f} = \frac{1}{W} \int_X f w \, d\mu$.

Proof By applying the result of Theorem 2.10 for the convex function $\varphi(x) - cx^2$, we get the desired result. The proof is complete. □

3.2 Refinements of Some Hermite-Hadamard-Type Inequalities for Strongly Convex Functions

As a first application of Theorem 3.2 we note the following: Particular choice of the measure space (X, μ) and the function f gives a continuous refinement of the left-hand side of the well-known Hermite-Hadamard inequality. In fact, by putting in (3.4) $X = [a, b]$, $d\mu(x) = \frac{1}{b-a} dx$, and $f(x) = x$ for $x \in [a, b]$, we get

$$\varphi\left(\frac{a+b}{2}\right) \leq \int_Z \varphi\left(\frac{1}{b-a} \int_a^b x \alpha(x, z) dx\right) d\lambda(z)$$

$$-c \int_Z \left(\frac{1}{b-a} \int_a^b x \alpha(x, z) dx - \frac{a+b}{2}\right)^2 d\lambda(z)$$

$$\leq \frac{1}{b-a} \int_a^b \varphi(x) \, dx - \frac{c}{12}(b-a)^2 \leq \frac{1}{b-a} \int_a^b \varphi(x) \, dx, \qquad (3.12)$$

where α satisfies (2.13) and (2.14).

The inequality between the first and the third term in the chain (3.12) is already known. It is the left-hand side of the Hermite-Hadamard inequality for a strongly convex function, and it was given in [79]. Hence, (3.12) is a continuous refinement of this result.

Discrete refinement of the left-hand side of the Hermite-Hadamard inequality for a convex function was given in [108]. Here we give a generalization of it, namely, a discrete refinement of the left-hand side of the Hermite-Hadamard inequality for a strongly convex function. It follows from (3.11) applied with $w(x) = 1$ and $f(x) = x$ for $x \in [a, b]$:

$$\varphi\left(\frac{a+b}{2}\right) \leq \frac{1}{b-a} \sum_{i=1}^n (a_i - a_{i-1}) \varphi\left(\frac{a_i + a_{i-1}}{2}\right)$$

$$-\frac{c}{b-a} \sum_{i=1}^n (a_i - a_{i-1}) \left(\frac{a_i + a_{i-1}}{2} - \frac{a+b}{2}\right)^2$$

$$\leq \frac{1}{b-a} \int_a^b \varphi(x)\,dx - \frac{c}{12}(b-a)^2 \leq \frac{1}{b-a}\int_a^b \varphi(x)\,dx. \qquad (3.13)$$

The particular case of (3.13) for $n = 2$, $a_0 = a$, $a_1 = \frac{a+b}{2}$, $a_2 = b$ was given in [10].

The refinement of the right-hand side of the Hermite-Hadamard inequality is based on the Lah-Ribarič inequality, and we cannot directly obtain a continuous refinement. A discrete refinement of the right-hand side of the Hermite-Hadamard inequality for a convex function was given in [108]. Here we derive a refinement of the Lah-Ribarič inequality for a strongly convex function which follows from the result for a convex function applied with the convex function $\varphi(x) - cx^2$ (see [102]).

Theorem 3.4 *Let f, w be integrable functions on $[a,b]$, $w \geq 0$, $W = \int_a^b w(t)\,dt \neq 0$, and let $a_0, a_1, \ldots, a_n$ be such that $a = a_0 < a_1 < \ldots < a_n = b$ and $m_i \leq f(t) \leq M_i$ for $t \in [a_{i-1}, a_i]$, $m_i \neq M_i$, $i = 1, 2, \ldots, n$ and $m = \min\{m_1, \ldots, m_n\}$, $M = \min\{M_1, \ldots, M_n\}$. If $\varphi : I \to \mathbb{R}$ is a strongly convex function with modulus c, $f([a,b]) \subseteq I$, then*

$$\frac{1}{W}\int_a^b \varphi(f(t))w(t)\,dt - \frac{c}{W}\int_a^b f^2(t)w(t)\,dt$$

$$\leq \frac{1}{W}\sum_{i=1}^n w_i\left[\frac{M_i - \bar{f}_i}{M_i - m_i}\varphi(m_i) + \frac{\bar{f}_i - m_i}{M_i - m_i}\varphi(M_i)\right]$$

$$-\frac{c}{W}\sum_{i=1}^n w_i[\bar{f}_i(M_i + m_i) - m_i M_i]$$

$$\leq \frac{M - \bar{f}}{M - m}\varphi(m) + \frac{\bar{f} - m}{M - m}\varphi(M) - c[\bar{f}(M + m) - mM], \qquad (3.14)$$

where $\bar{f} = \frac{1}{W}\int_a^b f(t)w(t)dt$, $w_i = \int_{a_{i-1}}^{a_i} w(t)dt$, and $\bar{f}_i = \frac{1}{w_i}\int_{a_{i-1}}^{a_i} f(t)w(t)dt$.
In particular, the following inequalities hold:

$$\frac{1}{b-a}\int_a^b \varphi(t)\,dt \leq \frac{1}{b-a}\sum_{i=1}^n (a_i - a_{i-1})\frac{\varphi(a_{i-1}) + \varphi(a_i)}{2}$$

$$-\frac{c}{b-a}\left[\sum_{i=1}^n (a_i - a_{i-1})\frac{a_i^2 + a_{i-1}^2}{2}\right] - \frac{c}{3}(a^2 + ab + b^2)$$

$$\leq \frac{\varphi(a) + \varphi(b)}{2} - c\frac{(b-a)^2}{6}. \qquad (3.15)$$

Proof If φ is a strongly convex function, then the function $\varphi(x) - cx^2$ is convex. Putting in Theorem 2.3 from paper [108] $\varphi(x) - cx^2$ instead of f, after simple calculations, we get (3.14).

Inequalities (3.15) follow from inequalities (3.14) with $w(t) = 1$ and $f(t) = t$. The proof is complete. □

As we can see, the chain of inequalities (3.14) is a refinement of the right-hand side of the Hermite-Hadamard inequality for a strongly convex function. If $n = 2$, $a_0 = a$, $a_1 = \frac{a+b}{2}$, $a_2 = b$, we get the result from [10, Theorem 5].

3.3 Refinements of the Jensen Inequality via Superquadracity

Results of this section involve superquadratic functions. First we recall some crucial information about the class of these functions.

Definition 3.2 A function $\varphi : [0, \infty) \to \mathbb{R}$ is superquadratic provided that for all $x \geq 0$ there exists a constant $C_x \in \mathbb{R}$ such that

$$\varphi(y) - \varphi(x) - \varphi(|y - x|) \geq C_x(y - x)$$

for all $y \geq 0$.

We say that f is subquadratic if $-f$ is superquadratic.

If φ is a superquadratic function with C_x as in definition, then it is known that (see [3]):

(a) $\varphi(0) \leq 0$.
(b) If $\varphi(0) = \varphi'(0) = 0$, then $C_x = \varphi'(x)$ whenever $\varphi'(x)$ exists.
(c) If $\varphi \geq 0$, then φ is convex and $\varphi(0) = \varphi'(0) = 0$.
(d) If φ is differentiable with $\varphi(0) = \varphi'(0) = 0$, then $\varphi'(x)/x^2$ is nondecreasing on $(0, \infty)$.

Moreover, if φ is continuously differentiable with $\varphi(0) \leq 0$ and if φ' is superadditive or $\varphi'(x)/x$ is nondecreasing, then φ is superquadratic (see [3]). Also, if φ is nonpositive, nonincreasing, superadditive function, then φ is superquadratic. Furthermore, a function $\varphi(x) = x^p$ is superquadratic if $p \geq 2$, and it is subquadratic for $1 < p < 2$.

We cite the following result, which is very useful in the proofs of our main results (see [3, Theorem 2.3] and c.f. also [4] and [120] for further details):

Theorem 3.5 *Let (Ω, μ) be a probability measure space. Then the inequality*

$$\varphi\left(\int_\Omega f(s)\,d\mu(s)\right)$$
$$\leq \int_\Omega \varphi(f(s))\,d\mu(s) - \int_\Omega \varphi\left(\left|f(s) - \int_\Omega f(s)\,d\mu(s)\right|\right)\,d\mu(s) \quad (3.16)$$

holds for all nonnegative μ-integrable functions f if and only if φ is superquadratic. Moreover, (3.16) holds in the reversed direction if and only if φ is subquadratic.

If φ is a nonnegative superquadratic function, then φ is convex (see [3, Lemma 2.2]) and inequality (3.16) is a refinement of the Jensen inequality for a convex function.

We need the following useful weighted version of Theorem 3.5:

Lemma 3.1 *Let φ be a superquadratic function and let t be a nonnegative measurable function such that $T = \int_\Omega t(s)\,ds > 0$. Then the inequality*

$$\varphi\left(\overline{f}\right) \leq \frac{1}{T}\int_\Omega t(s)\varphi(f(s))\,ds - \frac{1}{T}\int_\Omega t(s)\varphi\left(|f(s) - \overline{f}|\right)\,ds, \quad (3.17)$$

holds for all nonnegative functions f, where $\overline{f} = \frac{1}{T}\int_\Omega t(s)f(s)\,ds$. Moreover, (3.17) holds in the reversed direction if φ is subquadratic.

Proof Set $d\mu(s) = \frac{t(s)}{T}\,ds$. Then (3.17) follows from (3.16). The proof is complete. $\square$

Example 3.2 Let φ be a superquadratic function, and let x_1, x_2 be two nonnegative real numbers and $\lambda \in [0, 1]$. Then

$$\varphi(\lambda x_1 + (1-\lambda)x_2) \quad (3.18)$$
$$\leq \lambda\varphi(x_1) + (1-\lambda)\varphi(x_2) - \lambda\varphi\big((1-\lambda)|x_1 - x_2|\big) - (1-\lambda)\varphi\big(\lambda|x_1 - x_2|\big).$$

Moreover, (3.18) holds in the reversed direction if φ is subquadratic.

In fact, by taking $\Omega = [0, 1]$, $t(s) = 1$, and

$$f(s) = \begin{cases} x_1 : s \in [0, \lambda] \\ x_2 : s \in (\lambda, 1], \end{cases}$$

we see that (3.18) follows from (3.17).

3.3 Refinements of the Jensen Inequality via Superquadracity

Consider the nonnegative measurable functions α and β satisfying

$$\alpha(s) + \beta(s) = 1, \text{ for all } s \in \Omega.$$

Denote

$$T = \int_\Omega t(s)\,ds, \quad Q = \int_\Omega \alpha(s)t(s)\,ds \quad \text{and} \quad R = \int_\Omega \beta(s)t(s)\,ds.$$

Our first main result in this section was published in [103] and reads:

Theorem 3.6 *Let $\varphi : [0, \infty) \to \mathbb{R}$ be a superquadratic function, and let f be a nonnegative and measurable function. Then the following refined variant of Jensen-type inequality*

$$\varphi(\overline{f}) \leq \frac{Q}{T}\varphi(\overline{f}_Q) + \frac{R}{T}\varphi(\overline{f}_R) - \frac{Q}{T}\varphi\left(\frac{R}{T}|\overline{f}_Q - \overline{f}_R|\right) - \frac{R}{T}\varphi\left(\frac{Q}{T}|\overline{f}_Q - \overline{f}_R|\right)$$

$$= \frac{Q}{T}\varphi(\overline{f}_Q) + \frac{R}{T}\varphi(\overline{f}_R) - \frac{Q}{T}\varphi(|\overline{f}_Q - \overline{f}|) - \frac{R}{T}\varphi(|\overline{f}_R - \overline{f}|), \quad (3.19)$$

holds, where

$$\overline{f} = \frac{1}{T}\int_\Omega t(s)f(s)\,ds, \quad \overline{f}_Q = \frac{1}{Q}\int_\Omega \alpha(s)t(s)f(s)\,ds,$$

and

$$\overline{f}_R = \frac{1}{R}\int_\Omega \beta(s)t(s)f(s)\,ds.$$

Moreover, (3.19) holds in the reversed direction if φ is subquadratic.

Proof Set $x_1 = \overline{f}_Q$, $x_2 = \overline{f}_R$ and $\lambda = \frac{Q}{T}$. It is clear that

$$1 - \lambda = \frac{R}{T} \quad \text{and} \quad \lambda x_1 + (1-\lambda)x_2 = \overline{f}.$$

Then, from Example 3.2 it follows that

$$\varphi(\overline{f}) \leq \frac{Q}{T}\varphi(\overline{f}_Q) + \frac{R}{T}\varphi(\overline{f}_R) - \frac{Q}{T}\varphi\left(\frac{R}{T}|\overline{f}_Q - \overline{f}_R|\right) - \frac{R}{T}\varphi\left(\frac{Q}{T}|\overline{f}_Q - \overline{f}_R|\right).$$

Moreover, using the fact that

$$\frac{R}{T}|\overline{f}_Q - \overline{f}_R| = |\overline{f}_Q - \overline{f}| \quad \text{and} \quad \frac{Q}{T}|\overline{f}_Q - \overline{f}_R| = |\overline{f}_R - \overline{f}|,$$

we obtain inequality (3.19). The proof is complete, since the proof of the reversed inequality is similar to the proof above and we can omit the details. □

By making a further restriction of φ we can also state the following version of Theorem 3.6:

Theorem 3.7 *Let $\varphi : [0, \infty) \to \mathbb{R}$ be a nondecreasing and superquadratic function such that*

$$\varphi(a+b) \leq c\big(\varphi(a) + \varphi(b)\big), \text{ for some } c > 0. \tag{3.20}$$

Then the following refined variant of Jensen-type inequality

$$\varphi\left(\overline{f}\right) \leq I \leq \frac{1}{T}\int_\Omega t(s)\varphi(f(s))\,ds - \frac{1}{cT}\int_\Omega t(s)\varphi(|f(s) - \overline{f}|)\,ds$$

holds for all measurable functions f, where

$$I = \frac{Q}{T}\varphi\left(\overline{f}_Q\right) + \frac{R}{T}\varphi(\overline{f}_R) - \frac{Q}{T}\varphi\left(|\overline{f}_Q - \overline{f}|\right) - \frac{R}{T}\varphi(|\overline{f}_R - \overline{f}|).$$

Proof We proved the first inequality $\varphi\left(\overline{f}\right) \leq I$ in Theorem 3.6, so we only need to prove the second inequality.

By applying Lemma 3.1 in the first two terms of I, we get that

$$I \leq \frac{1}{T}\int_\Omega t(s)\varphi(f(s))\,ds - A_1 - B_1,$$

where

$$A_1 = \frac{Q}{T}\varphi\left(|\overline{f}_Q - \overline{f}|\right) + \frac{R}{T}\varphi(|\overline{f}_R - \overline{f}|)$$

and

$$B_1 = \frac{1}{T}\int_\Omega \alpha(s)t(s)\varphi(|f(s) - \overline{f}_Q|)\,ds + \frac{1}{T}\int_\Omega \beta(s)t(s)\varphi(|f(s) - \overline{f}_R|)\,ds.$$

To finish the proof, it is enough to prove that

$$\frac{1}{cT}\int_\Omega t(s)\varphi\big(|f(s)-\overline{f}|\big)\,ds \leq A_1 + B_1. \tag{3.21}$$

Now, by using the triangle inequality, nondecreasing property of φ and (3.20), we obtain that

$$\frac{1}{T}\int_\Omega \alpha(s)t(s)\varphi\big(|f(s)-\overline{f}|\big)\,ds$$
$$\leq c\left(\frac{1}{T}\int_\Omega \alpha(s)t(s)\varphi\big(|f(s)-\overline{f}_Q|\big)\,ds + \frac{Q}{T}\varphi\big(|\overline{f}_Q - \overline{f}|\big)\right)$$

and

$$\frac{1}{T}\int_\Omega \beta(s)t(s)\varphi\big(|f(s)-\overline{f}|\big)\,ds$$
$$\leq c\left(\frac{1}{T}\int_\Omega \beta(s)t(s)\varphi\big(|f(s)-\overline{f}_R|\big)\,ds + \frac{R}{T}\varphi\big(|\overline{f}_R - \overline{f}|\big)\right).$$

Hence (3.21) follows as a sum of the above two inequalities. The proof is complete. □

3.4 Refinements of the Continuous Minkowski and Beckenbach-Dresher Inequalities via Superquadracity

In the following discussion we consider the measurable spaces (X, μ), (X, λ), and (Y, ν). Moreover, $d\mu$, $d\lambda$ and $d\nu$ are notations for $d\mu(x)$, $d\lambda(x)$ and $d\nu(y)$, respectively. First, we remind about the following interesting refinement of the Hölder inequality by G. Sinnamon [120, Theorem 1.1]:

Lemma 3.2 *Let $p \geq 2$ and $\frac{1}{p} + \frac{1}{q} = 1$. Then*

$$\int_Y fg\,d\nu \leq \left(\int_Y (f^p - h^p)\,d\nu\right)^{\frac{1}{p}}\left(\int_Y g^q\,d\nu\right)^{\frac{1}{q}} \tag{3.22}$$

holds for any two nonnegative ν-measurable functions f and g, where

$$h = \left|f - \frac{g^{q-1}\int_Y fg\,d\nu}{\int_Y g^q\,d\nu}\right|.$$

Moreover, (3.22) holds in the reversed direction if $1 < p \leq 2$.

The natural "turning point" in Minkowski and Beckenbach-Dresher-type inequalities is $p = 1$, but in the refinements which involve a superquadratic function, the natural "turning point" is $p = 2$.

Our first main result in this section is the following refinement of the continuous Minkowski inequality (see [103]):

Theorem 3.8 *Let f be a nonnegative measurable function on $X \times Y$ with respect to the measure $\mu \times \nu$ and let $p \geq 2$. Then*

$$\left(\int_X \left(\int_Y f \, d\nu \right)^p d\mu \right)^{\frac{1}{p}} \leq \int_Y \left(\int_X (f^p - h^p) \, d\mu \right)^{\frac{1}{p}} d\nu, \qquad (3.23)$$

where

$$h = h(x, y) = \left| f - \frac{H \int_X f H^{p-1} \, d\mu}{\int_X H^p \, d\mu} \right|, \quad H(x) = \int_Y f(x, y) \, d\nu.$$

If $1 < p \leq 2$, then (3.23) holds in the reversed direction.

Proof Let $H(x) = \int_Y f(x, y) \, d\nu$. Let $p \geq 2$ and $\frac{1}{p} + \frac{1}{q} = 1$. Using Lemma 3.2, by replacing $f(x)$ and $g(x)$ with $f(x, y)$ and $H^{p-1}(x)$, respectively, we get that

$$\int_X f H^{p-1} \, d\mu \leq \left(\int_X (f^p - h^p) \, d\mu \right)^{\frac{1}{p}} \left(\int_X H^p \, d\mu \right)^{\frac{1}{q}}, \qquad (3.24)$$

where

$$h = h(x, y) = \left| f - \frac{H \int_X f H^{p-1} \, d\mu}{\int_X H^p \, d\mu} \right|.$$

We integrate inequality (3.24) over Y and apply the Fubini theorem on the left side of the inequality to find that

$$\left(\int_X H^p \, d\mu \right) \leq \left(\int_X H^p \, d\mu \right)^{\frac{1}{q}} \int_Y \left(\int_X (f^p - h^p) \, d\mu \right)^{\frac{1}{p}} d\nu. \qquad (3.25)$$

Since $H(x) = \int_Y f(x, y) \, d\nu$ and $1 - \frac{1}{q} = \frac{1}{p}$, we deduce that

$$\left(\int_X \left(\int_Y f \, d\nu \right)^p d\mu \right)^{\frac{1}{p}} \leq \int_Y \left(\int_X (f^p - h^p) \, d\mu \right)^{\frac{1}{p}} d\nu.$$

3.4 Refinements of the Continuous Minkowski and Beckenbach-...

The proof of the case $1 < p \leq 2$ is similar so we omit the details and the proof is complete. □

Next, we point out the following discrete version of Theorem 3.8:

Corollary 3.3 *Let $p \geq 2$ and let $f_1, f_2, \ldots, f_n$, $n = 2, 3, \ldots$, be nonnegative measurable functions. Then*

$$\left(\int_X \left(\sum_{i=1}^n f_i \right)^p d\mu \right)^{\frac{1}{p}} \leq \sum_{i=1}^n \left(\int_X (f_i^p - h_i^p) \, d\mu \right)^{\frac{1}{p}}, \tag{3.26}$$

where

$$H = H(x) = \sum_{i=1}^n f_i(x) \quad \text{and} \quad h_i = h_i(x, y) = \left| f_i - \frac{H \int_X f_i H^{p-1} d\mu}{\int_X H^p d\mu} \right|$$

for $i = 1, \ldots, n$.

If $1 < p \leq 2$, then (3.26) holds in the reversed direction.

Proof Let $Y = \bigcup_{i=1}^n Y_i$, where $Y_i = [i-1, i)$, for all $i = 1, \ldots, n$, and let $dv = dy$ be the Lebesgue measure.

Define $f(x, y) = \sum_{i=1}^n f_i(x) \chi_{Y_i}(y)$. Then

$$H(x) = \int_Y f(x, y) \, dv = \sum_{i=1}^n f_i(x)$$

and

$$h(x, y) = \left| f - \frac{H \int_X f H^{p-1} d\mu}{\int_X H^p d\mu} \right| = \sum_{i=1}^n \chi_{Y_i}(y) \left| f_i - \frac{H \int_X f_i H^{p-1} d\mu}{\int_X H^p d\mu} \right|$$

$$= \sum_{i=1}^n \chi_{Y_i}(y) h_i(x),$$

where

$$h_i = h_i(x, y) = \left| f_i - \frac{H \int_X f_i H^{p-1} d\mu}{\int_X H^p d\mu} \right|.$$

Therefore, by applying Theorem 3.8, one can complete the proof. □

Theorem 3.9 *Let p, s, and t be different real numbers such that $s \geq 2, t \geq 2$, and $(s-t)/(p-t) > 1$. Then for any positive measurable functions $f_1, \ldots, f_n$*

$$\left(\int_X \left(\sum_{i=1}^n f_i\right)^p d\mu\right)^{s-t}$$
$$\leq \left(\sum_{i=1}^n \left(\int_X (f_i^s - h_i^s) d\mu\right)^{\frac{1}{s}}\right)^{s\frac{p-t}{s-t}} \left(\sum_{i=1}^n \left(\int_X (f_i^t - r_i^t) d\mu\right)^{\frac{1}{t}}\right)^{t\frac{s-p}{s-t}}, \qquad (3.27)$$

where

$$H = \sum_{i=1}^n f_i, \quad h_i = h_i(x) = \left| f_i - \frac{H \int_X f_i H^{s-1} d\mu}{\int_X H^s d\nu} \right| \quad \text{and}$$

$$r_i = r_i(x) = \left| f_i - \frac{H \int_X f_i H^{t-1} d\mu}{\int_X H^t d\mu} \right|.$$

Moreover, if $p \neq 0$, $1 < t < 2$, $1 < s < 2$ and $(s-t)/(p-t) < 1$, then (3.27) holds in the reversed direction.

Proof Let $s \geq 2, t \geq 2$ such that $\frac{s-t}{p-t} > 1$. Putting in the integral Hölder inequality (1.2):

$$Y = X, \quad \nu = \mu, \quad p = \frac{s-t}{p-t}, \quad q = \frac{s-t}{s-p},$$

$$f = (f_1 + \ldots + f_n)^{\frac{s(p-t)}{s-t}} \quad \text{and} \quad g = (f_1 + \ldots + f_n)^{\frac{t(s-p)}{s-t}},$$

we get

$$\int_X \left(\sum_{i=1}^n f_i\right)^p d\mu = \int_X [(f_1 + \cdots + f_n)^s]^{\frac{p-t}{s-t}} [(f_1 + \cdots + f_n)^t]^{\frac{s-p}{s-t}} d\mu$$
$$\leq \left(\int_X (f_1 + \cdots + f_n)^s d\mu\right)^{\frac{p-t}{s-t}} \left(\int_X (f_1 + \cdots + f_n)^t d\mu\right)^{\frac{s-p}{s-t}}.$$

3.4 Refinements of the Continuous Minkowski and Beckenbach-...

Using Corollary 3.3, the above inequality becomes

$$\int_X \left(\sum_{i=1}^n f_i\right)^p d\mu$$

$$\leq \left(\sum_{i=1}^n \left(\int_X (f_i^s - h_i^s) d\mu\right)^{\frac{1}{s}}\right)^{s\frac{p-t}{s-t}} \left(\sum_{i=1}^n \left(\int_X (f_i^t - r_i^t) d\mu\right)^{\frac{1}{t}}\right)^{t\frac{s-p}{s-t}},$$

where

$$H = \sum_{i=1}^n f_i, \quad h_i = \left|f_i - \frac{H \int_X f_i H^{s-1} d\mu}{\int_X H^s d\mu}\right| \text{ and } r_i = \left|f_i - \frac{H \int_X f_i H^{t-1} d\mu}{\int_X H^t d\mu}\right|,$$

so the first case is proved.

The proof of the other case is similar so we omit the details and the proof is complete. □

Remark 3.2 Theorem 3.9 was given in [103], and it is both a generalization and refinement of Theorem 1.2 from [128].

Our new result related to the continuous Beckenbach-Dresher inequality reads (see [103]):

Theorem 3.10 *Let f and u be nonnegative measurable functions on $X \times Y$ with respect to the measures $\mu \times \nu$ and $\lambda \times \nu$, respectively, and let $1 < q \leq 2 \leq p, s \geq 1$. Then*

$$\frac{\left(\int_X \left(\int_Y f \, d\nu\right)^p d\mu\right)^{\frac{s}{p}}}{\left(\int_X \left(\int_Y u \, d\nu\right)^q d\lambda\right)^{\frac{s-1}{q}}} \leq \int_Y \frac{\left(\int_X (f^p - h^p) \, d\mu\right)^{\frac{s}{p}}}{\left(\int_X (u^q - r^q) \, d\lambda\right)^{\frac{s-1}{q}}} d\nu,$$

where

$$h = h(x, y) = \left|f - \frac{H \int_X f H^{p-1} d\mu}{\int_X H^p d\mu}\right|, \quad H = H(x) = \int_Y f(x, y) \, d\nu,$$

$$r = r(x, y) = \left|u - \frac{\hat{H} \int_X u \hat{H}^{q-1} d\mu}{\int_X \hat{H}^q d\mu}\right|, \quad \hat{H} = \hat{H}(x) = \int_Y u(x, y) \, d\nu.$$

Proof Let $1 < q \leq 2 \leq p$. Then, in view of Theorem 3.8 for $p \geq 2$ and $1 < q \leq 2$ we have that

$$\frac{\left(\int_X \left(\int_Y f \, dv\right)^p d\mu\right)^{\frac{s}{p}}}{\left(\int_X \left(\int_Y u \, dv\right)^q d\lambda\right)^{\frac{s-1}{q}}} \leq \frac{\left(\int_Y \left(\int_X (f^p - h^p) \, d\mu\right)^{\frac{1}{p}} dv\right)^s}{\left(\int_Y \left(\int_X (u^q - r^q) \, d\lambda\right)^{\frac{1}{q}} dv\right)^{s-1}}$$

$$= \left(\int_Y a^{\frac{1}{s}} dv\right)^s \left(\int_Y b^{\frac{1}{1-s}} dv\right)^{1-s}$$

$$\leq \int_Y ab \, dv,$$

where $a^{\frac{1}{s}} = \left(\int_X (f^p - h^p) \, d\mu\right)^{\frac{1}{p}}$ and $b^{\frac{1}{1-s}} = \left(\int_X (u^q - r^q) \, d\lambda\right)^{\frac{1}{q}}$. In the last inequality we used the reverse Hölder inequality for two functions a and b when one exponent $(1-s)$ is negative and the other exponent s is positive. The proof is complete. □

By using Theorem 3.10 and similar arguments as those in the proof of Corollary 3.3, we can also derive the following discrete version:

Corollary 3.4 Let $1 < q \leq 2 \leq p, s \geq 1$, $f_i, u_i : X \to [0, \infty)$ such that f_i^p, u_i^q are integrable, for all $i = 1, \ldots, n$. Then

$$\frac{\left(\int_X \left(\sum_{i=1}^n f_i\right)^p d\mu\right)^{\frac{s}{p}}}{\left(\int_X \left(\sum_{i=1}^n u_i\right)^q d\lambda\right)^{\frac{s-1}{q}}} \leq \sum_{i=1}^n \frac{\left(\int_X (f_i^p - h_i^p) \, d\mu\right)^{\frac{s}{p}}}{\left(\int_X (u_i^q - r_i^q) \, d\lambda\right)^{\frac{s-1}{q}}},$$

where

$$h_i = h_i(x) = \left| f_i - \frac{H \int_X f_i H^{p-1} d\mu}{\int_X H^p d\mu} \right|, \quad H = H(x) = \sum_{i=1}^n f_i(x),$$

$$r_i = r_i(x) = \left| u_i - \frac{\hat{H} \int_X u_i \hat{H}^{q-1} d\mu}{\int_X \hat{H}^q d\mu} \right|, \quad \hat{H} = \hat{H}(x) = \sum_{i=1}^n u_i(x).$$

As an application of Corollary 3.4, by making the substitution $s = \frac{p}{p-q}, p \neq q$, we obtain the following Beckenbach-Dresher-type inequality:

Example 3.3 Let $1 < q \leq 2 \leq p, q \neq p, f_i, u_i : X \to [0, \infty)$ such that f_i^p, u_i^q are integrable, for all $i = 1, \ldots, n$. Then

$$\left(\frac{\int_X \left(\sum_{i=1}^n f_i \right)^p d\mu}{\int_X \left(\sum_{i=1}^n u_i \right)^q d\lambda} \right)^{\frac{1}{p-q}} \leq \sum_{i=1}^n \left(\frac{\int_X (f_i^p - h_i^p) \, d\mu}{\int_X (u_i^q - r_i^q) \, d\lambda} \right)^{\frac{1}{p-q}},$$

where h_i and r_i are as in Corollary 3.4.

3.5 Refinements of the Continuous Hardy Inequality

The first refinement of the power weighted Hardy inequality by using superquadracity is the following (see [104]):

Theorem 3.11 *Let $p > 1$, $k > 1$, $0 < b \leq \infty$ and let f be a nonnegative integrable function on $(0, b)$ such that*

$$0 < \int_0^b x^{p-k} f^p(x) \, dx < \infty.$$

(i) If $p \geq 2$, then the following inequality holds:

$$\int_0^b x^{-k} \left(\int_0^x f(t) dt \right)^p dx + \frac{k-1}{p} \int_0^b \int_t^b \left| \frac{p}{k-1} \left(\frac{t}{x} \right)^{1-\frac{k-1}{p}} f(t) \right.$$

$$\left. - \frac{1}{x} \int_0^x f(t) \, dt \right|^p x^{p-k-\frac{k-1}{p}} t^{\frac{k-1}{p}-1} dx \, dt$$

$$\leq \left(\frac{p}{k-1} \right)^p \int_0^b x^{p-k} f^p(x) \left(1 - \left(\frac{x}{b} \right)^{\frac{k-1}{p}} \right) dx. \qquad (3.28)$$

(ii) If $1 < p \leq 2$, then (3.28) holds in the reversed direction.

The constant $\left(\frac{p}{k-1} \right)^p$ on the right-hand side of (3.28) is the best possible in all cases.

This result was generalized in a number of papers and especially in [5], (c.f. also [61]), where even the following continuous form was stated and proved with this technique by using superquadratic functions.

Theorem 3.12 *Let (X, μ) and (Y, ν) be measure spaces with positive σ-finite measures, u be a weight function on X, $k(x, y)$ a nonnegative measurable function on $X \times Y$, and $K(x) = \int_Y k(x, y) \, d\nu(y)$ for $x \in X$.*

Suppose that $K > 0$, that the function $x \mapsto u(x)\frac{k(x,y)}{K(x)}$ is integrable on X for each fixed $y \in Y$, and that v and H_k are defined on X by

$$v(y) = \int_X \frac{k(x, y)}{K(x)} u(x) \, d\mu(x) < \infty \tag{3.29}$$

and

$$H_k f(x) = \frac{1}{K(x)} \int_Y k(x, y) f(y) \, d\nu(y), \quad x \in X.$$

Suppose $I = (0, c)$, $c \leq \infty$, and $\varphi : I \to \mathbb{R}$. If φ is a superquadratic function, then the inequality

$$\int_X \varphi(H_k f(x)) u(x) \, d\mu(x)$$
$$+ \int_Y \int_X \varphi\Big(|f(y) - H_k f(x)|\Big) \frac{u(x) k(x, y)}{K(x)} \, d\mu(x) d\nu(y)$$
$$\leq \int_Y \varphi(f(y)) v(y) \, d\nu(y) \tag{3.30}$$

holds for all measurable functions $f : Y \to \mathbb{R}$, such that $\mathrm{Im} f \subseteq I$.

If φ is subquadratic, then the inequality sign in (3.30) is reversed.

Proof First we note that since $f(y) \in I$ for all $y \in Y$ (by assumption) and $H_k f(x)$ is an average also $H_k f(x) \in I$ for all $x \in X$. For a more formal proof of this fact, see also [61].

Now, let us prove inequality (3.30). By applying the refined Jensen inequality (3.8) to the first term on the left-hand side of (3.30) and then using Fubini theorem, we have that

$$\int_X \varphi(H_k f(x)) u(x) \, d\mu(x)$$
$$= \int_X u(x) \varphi\left(\frac{1}{K(x)} \int_Y k(x, y) f(y) d\nu(y)\right) d\mu(x)$$
$$\leq \int_X \frac{1}{K(x)} \left(\int_Y k(x, y) \varphi(f(y)) d\nu(y)\right) u(x) \, d\mu(x)$$
$$- \int_X \frac{u(x)}{K(x)} \int_Y k(x, y) \varphi(|f(y) - H_k f(x)|) \, d\nu(y) \, d\mu(x)$$
$$= \int_Y \varphi(f(y)) \left(\int_X \frac{k(x, y)}{K(x)} u(x) d\mu(x)\right) d\nu(y)$$

3.5 Refinements of the Continuous Hardy Inequality

$$-\int_Y \int_X \varphi\left(|f(y) - H_k f(x)|\right) \frac{u(x)k(x,y)}{K(x)} d\mu(x) \, d\nu(y)$$

$$= \int_Y \varphi(f(y))v(y) \, d\nu(y)$$

$$-\int_Y \int_X \varphi\left(|f(y) - H_k f(x)|\right) \frac{u(x)k(x,y)}{K(x)} d\mu(x) \, d\nu(y)$$

from which (3.30) follows.

If φ is subquadratic, then by making the same calculations as in the proof of (3.30), we see that only the inequality sign will be reversed. The proof is complete. □

Theorem 3.12 has a number of interesting special cases (for the proofs and more examples we refer to [5] and [61]).

Corollary 3.5 *Let the assumptions in Theorem* 3.12 *be satisfied.*

(i) If $p \geq 2$, then

$$\int_X H_k^p f(x) u(x) \, d\mu(x) + \int_Y \int_X |f(y) - H_k f(x)|^p \frac{u(x)k(x,y)}{K(x)} d\mu(x) \, d\nu(y)$$

$$\leq \int_Y f^p(y) v(y) \, d\nu(y). \qquad (3.31)$$

(ii) If $1 < p \leq 2$, then (3.31) *holds in the reversed direction.*

We also get the following refinement of the classical Hardy-Hilbert inequality (see [61]):

Corollary 3.6 *Let $p > 1$ and $f \in L_p(\mathbb{R}_+)$.*
If $p \geq 2$, then

$$\int_0^\infty \left(\int_0^\infty \frac{f(x)}{x+y} dx\right)^p dy$$

$$+ \left(\frac{\pi}{\sin \frac{\pi}{p}}\right)^{p-1} \int_0^\infty y^{-\frac{1}{p}} \int_0^\infty \left|f(y)y^{\frac{1}{p}} - \frac{\sin \frac{\pi}{p}}{\pi} x^{\frac{1}{p}} \int_0^\infty \frac{f(y)}{x+y} dy\right|^p \frac{x^{\frac{1}{p}-1}}{x+y} dx \, dy$$

$$\leq \left(\frac{\pi}{\sin \frac{\pi}{p}}\right)^p \int_0^\infty f^p(y) \, dy. \qquad (3.32)$$

If $1 < p \leq 2$, then (3.32) *holds in the reversed direction.*

4 Functionals Associated with Continuous Forms of Inequalities

In the classical case it is well-known that by studying some functionals related to an inequality one can obtain knowledge of independent interest but also provide an additional information about the inequality itself. One typical such functional is the so-called gap in the inequality—i.e., the difference of the right-hand and left-hand sides of the inequality, which typically gives even a refinement of the inequality itself. As an introductory example we mention the "Jensen gaps" studied in paper [1] (see also Remark 4.1). In this book, we have investigated some similar functionals (e.g., the "gaps") related to the continuous forms of inequalities.

4.1 Properties of Some Hölder-Type Functionals

In this section we fix the following objects: measure spaces (X, μ), (Y, ν), and (Z, λ), a positive function $f(x, y)$ measurable on $(X \times Y, \mu \times \nu)$, a nonnegative, and a measurable function $u(x)$ on X such that $\int_X u(x)\,d\mu(x) = 1$ and a nonnegative function $\alpha(z, y)$ on $Z \times Y$ such that

$$\int_Z \alpha(z, y)\,dz = 1, \quad \text{for } y \in Y.$$

By $\mathcal{K}_\mathcal{H}$ we denote the following set of weights on Y:

$$\mathcal{K}_\mathcal{H} = \{v : Y \to [0, \infty) : v \text{ is measurable}, \int_Y f(x, y)v(y)d\nu(y) > 0(\mu\text{-a.e.}),$$

$$\int_Y \alpha(z, y)f(x, y)v(y)d\nu(y) > 0(\mu\text{-a.e.})\}.$$

© The Author(s), under exclusive license to Springer Nature Switzerland AG 2025
L. Nikolova et al., *Continuous Versions of Some Classical Inequalities*, Frontiers in Mathematics, https://doi.org/10.1007/978-3-031-83372-4_4

Furthermore, we introduce the following abbreviations and notations:

$$L_H(v) = \int_Y \exp\left(\int_X \log f(x,y) u(x)\,d\mu(x)\right) v(y)\,d\nu(y),$$

$$M_H(v) = \int_Z \left[\exp \int_X \log\left(\int_Y \alpha(z,y) f(x,y) v(y)\,d\nu(y)\right) u(x)\,d\mu(x)\right] dz,$$

$$R_H(v) = \exp\left[\int_X \log\left(\int_Y f(x,y) v(y) d\nu(y)\right) u(x)\,d\mu(x)\right],$$

$$H_0(v) = R_H(v) - L_H(v),$$

$$H_1(v) = M_H(v) - L_H(v).$$

As we can see, the functional H_0 is the difference between the right-hand side and the left-hand side of the continuous Hölder inequality (1.4), while H_1 is the difference between the middle term in the refinement and the left-hand side of (2.3). In the following theorem some superadditivity properties of the functionals H_0, H_1, R_H and M_H are given (see [101]).

Theorem 4.1 *The functionals H_0, H_1, R_H and M_H are superadditive on $\mathcal{K}_H$, and L_H is linear.*

Proof Since the linearity of L_H is trivial, we have to prove that the following inequalities hold:

$$H_i(v+w) \geq H_i(v) + H_i(w), \quad i = 0, 1,$$

$$M_H(v+w) \geq M_H(v) + M_H(w),$$

$$R_H(v+w) \geq R_H(v) + R_H(w),$$

where v and w are weights from $\mathcal{K}_H$.

By putting in (2.7)

$$a(x) = \int_Y \alpha(z,y) f(x,y) v(y)\,d\nu(y), \quad b(x) = \int_Y \alpha(z,y) f(x,y) w(y)\,d\nu(y),$$

we get that

$$\exp\left[\int_X \log\left(\int_Y \alpha(z,y) f(x,y) v(y)\,d\nu(y)\right) u(x)\,d\mu(x)\right]$$
$$+ \exp\left[\int_X \log\left(\int_Y \alpha(z,y) f(x,y) w(y)\,d\nu(y)\right) u(x)\,d\mu(x)\right]$$

$$\leq \exp\left[\int_X \log\left(\int_Y \alpha(z,y)f(x,y)(v(y)+w(y))\,d\nu(y)\right)u(x)\,d\mu(x)\right].$$

Now by integrating over Z, we obtain that $M_H(v+w) \geq M_H(v) + M_H(w)$. The superadditivity of R_H can be proved in the similar manner. In our consideration of H_i we use the fact that $L_H(v+w) = L_H(v) + L_H(w)$ and the just obtained properties for M_H and R_H. In particular, for H_1 we have

$$H_1(v+w) - H_1(v) - H_1(w)$$

$$= \Big(M_H(v+w) - M_H(v) - M_H(w)\Big) - \Big(L_H(v+w) - L_H(v) - L_H(w)\Big) \geq 0.$$

The proof of the superadditivity of H_0 is similar and thus omitted, so the proof is complete. □

Corollary 4.1 *The functionals H_0, H_1, R_H and M_H are nondecreasing.*

Proof This is a consequence of the positivity and superadditivity of the considered functionals. For example, let us prove this fact for the functional H_0. If $v \geq w$, then $v - w \geq 0$, and from Theorem 1.4 we get that $H_0(v-w) \geq 0$. Hence, by using Theorem 4.1, we obtain that

$$H_0(v) = H_0(w + (v-w)) \geq H_0(w) + H_0(v-w) \geq H_0(w).$$

The proofs for the other functionals are similar and therefore omitted. The proof is complete. □

The following text in this section is devoted to functionals related to the Popoviciu inequality (see [101]). Let (X, μ), (Y, ν), (Z, λ) be measure spaces and the functions $f(x,y)$, $\alpha(z,y)$, $u(x)$, and $f_0(x)$ be fixed and satisfy the assumptions of Theorem 2.5. With $\mathcal{K}_\mathcal{P}$ we denote a set of pairs (v_0, v) such that $v_0 \in (0, \infty)$, v is a weight on Y, and $v_0 f_0(x) - \int_Y f(x,y)v(y)\,d\nu(y) > 0$ for $x \in X$.

We define C_H and K_H as follows:

$$C_H(v_0) = \exp\left(\int_X \log\Big(v_0 f_0(x)\Big)u(x)\,d\mu(x)\right),$$

$$K_H(v_0, v) = \exp\left[\int_X \log\left(v_0 f_0(x) - \int_Y f(x,y)v(y)\,d\nu(y)\right)u(x)\,d\mu(x)\right].$$

We also define the following functionals:

$$P_1(v_0, v) = C_H(v_0) - L_H(v) - K_H(v_0, v),$$

$$P_2(v_0, v) = C_H(v_0) - R_H(v) - K_H(v_0, v),$$

$$P_3(v_0, v) = C_H(v_0) - M_H(v) - K_H(v_0, v).$$

Using these definitions, the continuous Popoviciu inequality (Theorem 1.7) has the following form:

$$C_H(v_0) - L_H(v) \geq K_H(v_0, v),$$

i.e., $P_1(v_0, v) \geq 0$.

Some properties of these functionals are given in the following theorem.

Theorem 4.2 *If $(v_0, v), (w_0, w) \in \mathcal{K}_\mathcal{P}$, then $C_H(v_0)$ is a linear function, and*

$$K_H(v_0 + w_0, v + w) \geq K_H(v_0, v) + K_H(w_0, w), \tag{4.1}$$

$$P_i(v_0 + w_0, v + w) \leq P_i(v_0, v) + P_i(w_0, w), \quad i = 1, 2, 3. \tag{4.2}$$

Proof Since $C_H(v_0)$ is equal to $v_0 \exp\left(\int_X \log f_0(x) u(x) \, d\mu(x)\right)$, it is obvious that C_H is linear. Let us prove inequality (4.1) for the functional K_H. By using (2.7) with

$$a(x) = v_0 f_0^p(x) - \int_Y f^p(x, y) v(y) \, d\nu(y)$$

and

$$b(x) = w_0 f_0^p(x) - \int_Y f^p(x, y) w(y) \, d\nu(y),$$

we obtain (4.1).

For the functional P_1 we have the following:

$$P_1(v_0 + w_0, v + w) - P_1(v_0, v) - P_1(w_0, w)$$

$$= \Big[C_H(v_0 + w_0) - C_H(v_0) - C_H(w_0)\Big] - \Big[L_H(v + w) - L_H(v) - L_H(w)\Big]$$

$$- \Big[K_H(v_0 + w_0, v + w) - K_H(v_0, v) - K_H(w_0, w)\Big]$$

$$= -\Big[K_H(v_0 + w_0, v + w) - K_H(v_0, v) - K_H(w_0, w)\Big] \leq 0,$$

since C_H and L_H are linear and K_H satisfies (4.1). Hence, (4.2) is proved for P_1.

Moreover, inequality (4.2) for P_2 and P_3 follows easily from the already proved properties of C_H, M_H, R_H and K_H so we omit the details. The proof is complete. □

4.1 Properties of Some Hölder-Type Functionals

In Sect. 2.1 we considered a continuous refinement of the Hölder inequality. Let us write the inequality which follows from Corollary 2.1 and Remark 2.1. Namely, let $p_1, \ldots, p_n$ be nonnegative real numbers such that $\sum_{i=1}^{n} p_i = 1$, and $f_1, \ldots, f_n$ be nonnegative functions on Y such that $f_i, \alpha(z, .) f_i \in L_{p_i}(Y)$, $i = 1, \ldots, n$, $\prod_{i=1}^{n} f_i^{p_i} \in L_1(Y)$, where α satisfies assumptions of Theorem 2.3. Then, by putting in (2.4): $X = X_1 \cup \ldots \cup X_n$, $X_1, \ldots, X_n$ are mutually disjoint, $\int_{X_i} u(x) \, d\mu(x) = p_i$, $p(x) = p_i$ for $x \in X_i$ and $f(x, y) = f_i(y)$ for $x \in X_i$, we get

$$\int_Y \prod_{i=1}^{n} f_i^{p_i}(y) v(y) d\nu(y) \leq \int_Z \prod_{i=1}^{n} \left(\int_Y \alpha(z, y) f_i(y) v(y) d\nu(y) \right)^{p_i} dz$$

$$\leq \prod_{i=1}^{n} \left(\int_Y f_i(y) v(y) d\nu(y) \right)^{p_i}. \quad (4.3)$$

Now we will study two functionals involving terms which appear in inequality (4.3).

Let $f_1, \ldots, f_n$ be nonnegative measurable functions on Y, v be a weight on Y, and α be a function, which satisfies assumptions of Theorem 2.3. By $T_{\mathbf{f}}$ we denote a set of nonnegative n-tuples $\mathbf{r} = (r_1, \ldots, r_n)$ with $R = \sum_{i=1}^{n} r_i > 0$ such that the integrals $\int_Y \prod_{i=1}^{n} f_i^{r_i/R}(y) v(y) d\nu(y)$ and $\int_Z \prod_{i=1}^{n} \left(\int_Y \alpha(z, y) f_i(y) v(y) d\nu(y) \right)^{r_i/R} dz$ are positive.
On the set $T_{\mathbf{f}}$ we define functionals D_1 and D_2 as follows:

$$D_1(\mathbf{r}) = \frac{\prod_{i=1}^{n} \left(\int_Y f_i(y) v(y) d\nu(y) \right)^{r_i}}{\left[\int_Y \prod_{i=1}^{n} f_i^{r_i/R}(y) v(y) d\nu(y) \right]^R},$$

$$D_2(\mathbf{r}) = \frac{\prod_{i=1}^{n} \left(\int_Y f_i(y) v(y) d\nu(y) \right)^{r_i}}{\left[\int_Z \prod_{i=1}^{n} \left(\int_Y \alpha(z, y) f_i(y) v(y) d\nu(y) \right)^{r_i/R} dz \right]^R}.$$

Since $\sum_{i=1}^{n} \frac{r_i}{R} = 1$, inequality (4.3) yields that $D_1(\mathbf{r}) \geq 1$ and $D_2(\mathbf{r}) \geq 1$. The following theorem states in particular that D_1 and D_2 have the following properties:

Theorem 4.3 *If $\mathbf{r}, \mathbf{s} \in T_{\mathbf{f}}$, then*

$$D_i(\mathbf{r} + \mathbf{s}) \geq D_i(\mathbf{r}) \cdot D_i(\mathbf{s}), \quad i = 1, 2. \quad (4.4)$$

Moreover, if $\mathbf{r} \geq \mathbf{s}$ *such that* $\mathbf{r} - \mathbf{s} \in T_{\mathbf{f}}$, *then*

$$D_i(\mathbf{r}) \geq D_i(\mathbf{s}), \quad i = 1, 2.$$

If m and M are positive numbers such that $m\mathbf{s} \leq \mathbf{r} \leq M\mathbf{s}$, *then*

$$D_i^m(\mathbf{s}) \leq D_i(\mathbf{r}) \leq D_i^M(\mathbf{s}), \quad i = 1, 2.$$

Proof We will first prove (4.4) for D_2. Let us transform the denominator of $D_2(\mathbf{r} + \mathbf{s})$:

$$\left[\int_Z \prod_{i=1}^n \left(\int_Y \alpha f_i v \, dv \right)^{\frac{r_i+s_i}{R+S}} dz \right]^{R+S}$$

$$= \left[\int_Z \left(\prod_{i=1}^n \left(\int_Y \alpha f_i v \, dv \right)^{\frac{r_i}{R}} \right)^{\frac{R}{R+S}} \left(\prod_{i=1}^n \left(\int_Y \alpha f_i v \, dv \right)^{\frac{s_i}{S}} \right)^{\frac{S}{R+S}} dz \right]^{R+S}$$

$$\leq \left[\int_Z \prod_{i=1}^n \left(\int_Y \alpha f_i v \, dv \right)^{\frac{r_i}{R}} dz \right]^R \left[\int_Z \prod_{i=1}^n \left(\int_Y \alpha f_i v \, dv \right)^{\frac{s_i}{S}} dz \right]^S, \qquad (4.5)$$

where in the last inequality we use the Hölder inequality (4.3) applied on two functions with exponents $p_1 = \frac{R}{R+S}$ and $p_2 = \frac{S}{R+S}$. From (4.5) and since

$$\prod_{i=1}^n \left(\int_Y f_i v \, dv \right)^{r_i+s_i} = \prod_{i=1}^n \left(\int_Y f_i v \, dv \right)^{r_i} \cdot \prod_{i=1}^n \left(\int_Y f_i v \, dv \right)^{s_i},$$

we get that (4.4) holds for D_2. The proof of (4.4) for D_1 can be done in a similar way.

If $\mathbf{r} \geq \mathbf{s}$, then, since $D_i(\mathbf{r} - \mathbf{s}) \geq 1$, we obtain from (4.4) that

$$D_i(\mathbf{r}) = D_i((\mathbf{r} - \mathbf{s}) + \mathbf{s}) \geq D_i(\mathbf{r} - \mathbf{s}) D_i(\mathbf{s}) \geq D_i(\mathbf{s}),$$

i.e., both D_1 and D_2 are increasing. The last inequality in Theorem 4.3 follows because $D_i(m\mathbf{r}) = D_i^m(\mathbf{r})$ and $D_i(M\mathbf{r}) = D_i^M(\mathbf{r})$ and D_i is nondecreasing ($i = 1, 2$). The proof is complete. □

As a consequence of Theorem 4.3 we have the following refinement of the continuous Hölder inequality:

Theorem 4.4 *Let* $n = 2, 3, \ldots$ *and* $\mathbf{r} = (r_1, \ldots, r_n)$. *If* $\mathbf{r}, (\frac{1}{n}, \ldots, \frac{1}{n}) \in T_{\mathbf{f}}$, *then the following refinements of the continuous Hölder inequality hold:*

$$\left[\frac{\prod_{i=1}^{n}\int_{Y}f_{i}vdv}{\left[\int_{Y}\prod_{i=1}^{n}f_{i}^{1/n}vdv\right]^{n}}\right]^{r_{min}} \leq \frac{\prod_{i=1}^{n}\left(\int_{Y}f_{i}vdv\right)^{r_{i}}}{\left[\int_{Y}\prod_{i=1}^{n}f_{i}^{r_{i}/R}vdv\right]^{R}} \leq \left[\frac{\prod_{i=1}^{n}\int_{Y}f_{i}vdv}{\left[\int_{Y}\prod_{i=1}^{n}f_{i}^{1/n}vdv\right]^{n}}\right]^{r_{max}}$$

and

$$\left[\frac{\prod_{i=1}^{n}\int_{Y}f_{i}vdv}{\left[\int_{Z}\prod_{i=1}^{n}\left(\int_{Y}\alpha f_{i}vdv\right)^{1/n}dz\right]^{n}}\right]^{r_{min}}$$

$$\leq \frac{\prod_{i=1}^{n}\left(\int_{Y}f_{i}vdv\right)^{r_{i}}}{\left[\int_{Z}\prod_{i=1}^{n}\left(\int_{Y}\alpha f_{i}vdv\right)^{r_{i}/R}dz\right]^{R}} \leq \left[\frac{\prod_{i=1}^{n}\int_{Y}f_{i}vdv}{\left[\int_{Z}\prod_{i=1}^{n}\left(\int_{Y}\alpha f_{i}vdv\right)^{1/n}dz\right]^{n}}\right]^{r_{max}},$$

where $r_{min} = \min\{r_{1},\ldots r_{n}\}$ and $r_{max} = \max\{r_{1},\ldots r_{n}\}$.

Proof We consider the n-tuples

$$\mathbf{s_1} = (r_{min},\ldots,r_{min}), \quad \mathbf{s_2} = (r_{max},\ldots,r_{max}).$$

Obviously $\mathbf{s_1} \leq \mathbf{r} \leq \mathbf{s_2}$, and both inequalities follow from Theorem 4.3. The proof is complete. □

4.2 Properties of Some Minkowski-Type Functionals

In this section we work with fixed measure spaces (X, μ), (Y, ν), (Z, λ), number $p \geq 1$, functions $f(x, y), \alpha(z, y), u(x)$, and $f_0(x)$ which satisfy the assumptions of Theorems 2.6 and 2.7. With $\mathcal{K}_M$ we denote a set of weights v on Y, and with $\mathcal{K}_B$ we denote a set of pairs (v_0, v) such that $v_0 \in (0, \infty)$, v is a weight on Y, and $v_0 f_0^p(x) - \int_Y f^p(x,y)v(y)dv(y) > 0$ for $x \in X$.

We define the following functionals:

$$L_M(v) = \int_Y \left(\int_X f(x,y)u(x)\,d\mu(x)\right)^p v(y)\,dv(y),$$

$$M_M(v) = \int_Z \left[\int_X \left(\int_Y \alpha(z, y) f^p(x, y) v(y) \, dv(y) \right)^{1/p} u(x) \, d\mu(x) \right]^p dz,$$

$$R_M(v) = \left[\int_X \left(\int_Y f^p(x, y) v(y) \, dv(y) \right)^{1/p} u(x) \, d\mu(x) \right]^p.$$

Using these definitions the refinement of the Minkowski inequality in Theorem 2.6 has the following form:

$$L_M \leq M_M \leq R_M.$$

In our next theorem some properties of the functionals L_M, R_M, M_M, $M_0 = R_M - L_M$, and $M_1 = M_M - L_M$ are given (see [52, 101]).

Theorem 4.5 *The functional L_M is linear on $\mathcal{K}_M$, and for $v, w \in \mathcal{K}_M$ the functionals M_0, M_1, R_M and M_M satisfy the following properties:*

$$M_i(v + w) \geq M_i(v) + M_i(w), \quad i = 0, 1,$$

$$M_M(v + w) \geq M_M(v) + M_M(w),$$

$$R_M(v + w) \geq R_M(v) + R_M(w).$$

Proof Let us first recall the following well-known reversed form of integral Minkowski inequality for two functions and exponent $p \geq 1$ (see (1.13)):

$$\left[\int_X \left(a^p(x) + b^p(x) \right)^{\frac{1}{p}} u(x) \, d\mu(x) \right]^p$$
$$\geq \left[\int_X a(x) u(x) \, d\mu(x) \right]^p + \left[\int_X b(x) u(x) \, d\mu(x) \right]^p, \quad (4.6)$$

where $a(x) \geq 0$ and $b(x) \geq 0$.

By using (4.6) with

$$a(x) = \left(\int_Y f^p(x, y) v(y) \, dv(y) \right)^{\frac{1}{p}} \quad \text{and} \quad b(x) = \left(\int_Y f^p(x, y) w(y) \, dv(y) \right)^{\frac{1}{p}},$$

we get that

$$\left[\int_X \left(\int_Y f^p(x, y) v(y) \, dv(y) + \int_Y f^p(x, y) w(y) \, dv(y) \right)^{\frac{1}{p}} u(x) \, d\mu(x) \right]^p$$

4.2 Properties of Some Minkowski-Type Functionals

$$\geq \left[\int_X \left(\int_Y f^p(x,y) v(y) \, d\nu(y) \right)^{\frac{1}{p}} u(x) \, d\mu(x) \right]^p$$
$$+ \left[\int_X \left(\int_Y f^p(x,y) w(y) \, d\nu(y) \right)^{\frac{1}{p}} u(x) \, d\mu(x) \right]^p,$$

i.e.,

$$R_M(v+w) \geq R_M(v) + R_M(w).$$

The proof of the corresponding inequality for M_M can be done in the similar way, so we omit the details. Moreover, the stated properties of M_0 and M_1 follow from the properties of L_M, M_M, and R_M. The proof is complete. □

Corollary 4.2 *The functionals M_0, M_1, R_M and M_M are nondecreasing.*

Proof We prove the statement for functional M_1. If $v \geq w$, then $v - w \geq 0$, and from Theorem 1.2, we get that $M_1(v - w) \geq 0$. Hence, by using Theorem 4.5 we obtain that

$$M_1(v) = M_1(w + (v-w)) \geq M_1(w) + M_1(v-w) \geq M_1(w).$$

Proofs for the other functionals M_0, R_M and M_M can be done in the similar way. The proof is complete. □

In our next discussion about properties of functionals related to the refinement of the Bellman inequality, besides the functionals L_M, R_M and M_M, we introduce the functionals C_M, K_M, B_1, B_2 and B_3, defined as follows:

$$C_M(v_0) = v_0 \left[\int_X f_0(x) u(x) \, d\mu(x) \right]^p,$$

$$K_M(v_0, v) = \left(\int_X \left[v_0 f_0^p(x) - \int_Y f^p(x,y) v(y) \, d\nu(y) \right]^{\frac{1}{p}} u(x) \, d\mu(x) \right)^p,$$

and

$$B_1(v_0, v) = C_M(v_0) - L_M(v) - K_M(v_0, v),$$
$$B_2(v_0, v) = C_M(v_0) - R_M(v) - K_M(v_0, v),$$
$$B_3(v_0, v) = C_M(v_0) - M_M(v) - K_M(v_0, v).$$

Theorem 4.6 *The functionals C_M and L_M are linear on $\mathcal{K}_M$, and for (v_0, v), (w_0, w) from $\mathcal{K}_B$ the following holds:*

$$K_M(v_0 + w_0, v + w) \geq K_M(v_0, v) + K_M(w_0, w),$$

$$B_i(v_0 + w_0, v + w) \leq B_i(v_0, v) + B_i(w_0, w), \quad i = 1, 2, 3.$$

Proof The proofs of the linearity of C_M and L_M are obvious. Next we prove the stated inequality for the functional K_M. By inserting

$$a(x) = \left[v_0 f_0^p(x) - \int_Y f^p(x, y) v(y) \, dv(y) \right]^{\frac{1}{p}}$$

$$b(x) = \left[w_0 f_0^p(x) - \int_Y f^p(x, y) w(y) \, dv(y) \right]^{\frac{1}{p}}$$

into (4.6), we get

$$a^p(x) + b^p(x) = (v_0 + w_0) f_0^p(x) - \int_Y f^p(x, y)(v(y) + w(y)) \, dv(y),$$

and the inequality for K_M follows from (4.6). The stated inequalities concerning B_i, $i = 1, 2, 3$, follow from the above-mentioned properties of C_M, L_M, and K_M and the superadditivity of M_M and R_M. The proof is complete. □

4.3 Properties of Some Jensen-Type Functionals

In this section we study properties of functionals related to inequalities of continuous Jensen-type, which were proved in Sect. 2.3 and based on results from [102]. We fix the following objects: a measure space (X, μ), a probability measure space (Z, λ), a convex function $\varphi : I \to \mathbb{R}$, functions $f : X \to I$, $\alpha : X \times Z \to [0, \infty)$ such that $\int_Z \alpha(x, z) \, d\lambda(z) = 1$ ($x \in X$), and a positive number f_0.

By $\mathcal{K}_{\varphi, f, \alpha}$ we denote the set of weights $w : X \to [0, \infty)$ such that $wf, w(\varphi \circ f) \in L_1(X)$ and $\int_X \alpha w \, d\mu = \int_X w \, d\mu \neq 0$.

By $\mathcal{K}_{\varphi, f_0, f, \alpha}$ we denote a class of pairs (w_0, w) such that $w_0 \in (0, \infty)$, $w : X \to [0, \infty)$, $wf, w(\varphi \circ f) \in L_1(X)$, $\int_X \alpha w \, d\mu = \int_X w \, d\mu \neq 0$, $w_0 - \int_X w \, d\mu > 0$,

$$\frac{w_0 f_0 - \int_X f w \, d\mu}{w_0 - \int_X w \, d\mu} \in I.$$

4.3 Properties of Some Jensen-Type Functionals

We define the functionals L_J, M_J, R_J, K_J as follows:

$$L_J(w) = W \cdot \varphi\left(\frac{1}{W}\int_X fw\,d\mu\right)$$

$$M_J(w) = W \cdot \int_Z \varphi\left(\frac{1}{W}\int_X f(x)\alpha(x,z)w(x)\,d\mu(x)\right) d\lambda(z)$$

$$R_J(w) = \int_X (\varphi \circ f)w\,d\mu$$

$$K_J(w_0, w) = (w_0 - W) \cdot \varphi\left(\frac{w_0 f_0 - \int_X fw\,d\mu}{w_0 - W}\right).$$

Here, as usual, $W = \int_X w\,d\mu$. Moreover,

$$J_1(w) = R_J(w) - L_J(w)$$
$$J_2(w) = R_J(w) - M_J(w).$$

First we state the following complementary information about the "gaps" in the inequalities (2.15).

Theorem 4.7 *The functionals L_J and M_J are subadditive on $K_{\varphi,f,\alpha}$, i.e.,*

$$L_J(v+w) \leq L_J(v) + L_J(w)$$
$$M_J(v+w) \leq M_J(v) + M_J(w)$$

for all $v, w \in K_{\varphi,f,\alpha}$ and the functionals J_1 and J_2 are superadditive. Moreover, if $p, q \in K_{\varphi,f,\alpha}$ satisfy $p \leq q$, then

$$J_1(p) \leq J_1(q) \quad and \quad J_2(p) \leq J_2(q).$$

Proof Let us denote: $S(w) = W \cdot \varphi\left(\frac{1}{W}\int_X f(x)\alpha(x,z)w(x)\,d\mu(x)\right)$. Since φ is a convex function, we get

$$S(v+w) = (V+W)\varphi\left(\frac{1}{V+W}\int_X (v+w)f\alpha\,d\mu\right)$$
$$= (V+W)\varphi\left(\frac{V}{V+W}\left(\frac{1}{V}\int_X vf\alpha\,d\mu\right) + \frac{W}{V+W}\left(\frac{1}{W}\int_X wf\alpha\,d\mu\right)\right)$$
$$\leq (V+W) \cdot \left[\frac{V}{V+W}\varphi\left(\frac{1}{V}\int_X vf\alpha\,d\mu\right) + \frac{W}{V+W}\varphi\left(\frac{1}{W}\int_X wf\alpha\,d\mu\right)\right]$$

$$= V \cdot \varphi \left(\frac{1}{V} \int_X v f \alpha \, d\mu \right) + W \cdot \varphi \left(\frac{1}{W} \int_X w f \alpha \, d\mu \right)$$
$$= S(v) + S(w),$$

where $V = \int_X v \, d\mu$ and $W = \int_X w \, d\mu$. In other words, we get that $S(v+w) \leq S(v) + S(w)$. By integrating the terms in this inequality over Z, we get the subadditivity of the functional M_J. Using the same method, we can prove the subadditivity of the functional L_J.

Since L_J and M_J are subadditive and R_J is linear, we obtain that J_1 and J_2 are superadditive.

By using the results in Sect. 2.3, we see that the functionals J_1 and J_2 are nonnegative on $K_{\varphi,f,\alpha}$. If $p \leq q$, then from the superadditivity of J_i, $i = 1, 2$, we get

$$J_i(q) = J_i(p + (q-p)) \geq J_i(p) + J_i(q-p) \geq J_i(p),$$

i.e., $J_i, i = 1, 2$, are nondecreasing functionals. The proof is complete. □

Now we define the following functionals C, A_1, A_2, and A_3 (connected to the reverse Jensen inequality):

$$C(w_0) = w_0 \varphi(f_0),$$
$$A_1(w_0, w) = C(w_0) - R_J(w) - K_J(w_0, w),$$
$$A_2(w_0, w) = C(w_0) - L_J(w) - K_J(w_0, w),$$
$$A_3(w_0, w) = C(w_0) - M_J(w) - K_J(w_0, w).$$

Our corresponding results for the "gaps" in inequality (2.19) read:

Theorem 4.8 *The functionals A_1, A_2, A_3 are superadditive on $K_{\varphi, f_0, f, \alpha}$, i.e.,*

$$A_i(v_0 + w_0, v + w) \geq A_i(v_0, v) + A_i(w_0, w), \quad i = 1, 2, 3,$$

for all $(v_0, v), (w_0, w) \in K_{\varphi, f_0, f, \alpha}$. Moreover,

$$A_1 \leq A_3 \leq A_2 \leq 0. \tag{4.7}$$

Proof First we prove the property of K_J. Putting in the definition of the convex function φ:

$$(r+s)\varphi \left(\frac{rx + sy}{r+s} \right) \leq r\varphi(x) + s\varphi(y)$$

4.3 Properties of Some Jensen-Type Functionals

the following substitutions:

$$r = v_0 - V, \quad s = w_0 - W, \quad \text{where } V = \int_X v \, d\mu, \quad W = \int_X w \, d\mu,$$

$$x = \frac{v_0 f_0 - \int_X v f \, d\mu}{v_0 - V} \quad \text{and} \quad y = \frac{w_0 f_0 - \int_X w f \, d\mu}{w_0 - W},$$

we get that

$$K_J(v_0 + w_0, v + w)$$
$$= ((v_0 + w_0) - (V + W)) \cdot \varphi \left(\frac{(v_0 f_0 + w_0 f_0) - \int_X (v + w) f \, d\mu}{(v_0 + w_0) - (V + W)} \right)$$
$$\leq (v_0 - V) \cdot \varphi \left(\frac{v_0 f_0 - \int_X v f \, d\mu}{v_0 - V} \right) + (w_0 - W) \cdot \varphi \left(\frac{w_0 f_0 - \int_X w f \, d\mu}{w_0 - W} \right)$$
$$= K_J(v_0, w) + K_J(v_0, w).$$

Hence, the subadditivity of K_J is proved. From this fact and from the linearity of C and R_J the superadditivity of A_1 also follows. The superadditivity of A_2 and A_3 can be proved in similar ways, so we omit the details.

The inequalities in (4.7) follow from the refinement of the Jensen inequality (see Theorem 2.8) and from (2.10). Indeed, the inequalities $A_1 \leq A_3 \leq A_2$ follow from (2.15) and $A_2 \leq 0$ holds in view of (2.17). The proof is complete. □

Next we consider functionals where the variable is a convex function. We refer to notations and results of Sect. 2.3. Let w, f and α be fixed functions. We define two linear functionals $\mathcal{J}_1$ and $\mathcal{J}_2$ as follows:

$$\mathcal{J}_1(\varphi) = \int_X (\varphi \circ f) w \, d\mu - W \cdot \varphi \left(\frac{1}{W} \int_X f w \, d\mu \right),$$

$$\mathcal{J}_2(\varphi) = \int_X (\varphi \circ f) w \, d\mu - W \int_Z \varphi \left(\frac{1}{W} \int_X f(x) \alpha(x, z) w(x) \, d\mu(x) \right) d\lambda(z),$$

provided that all integrals exist.

Theorem 4.9 *Let $i \in \{1, 2\}$. If $\varphi \in C^2([a, b])$, then there exists $c_i \in [a, b]$ such that*

$$\mathcal{J}_i(\varphi) = \frac{1}{2} \varphi''(c_i) \mathcal{J}_i(\varphi_0),$$

where $\varphi_0(x) = x^2$.

Proof Let us fix $i \in \{1, 2\}$. Since φ'' is continuous on segment $[a, b]$, then φ'' has a minimum and a maximum, i.e., there exist numbers m and M such that

$$m = \min_{x \in [a,b]} \varphi''(x) \quad \text{and} \quad M = \max_{x \in [a,b]} \varphi''(x).$$

Thus, functions

$$\varphi_1(x) = \frac{M}{2}x^2 - \varphi(x) \quad \text{and} \quad \varphi_2(x) = \varphi(x) - \frac{m}{2}x^2$$

are convex. By Theorem 2.8, we get

$$\mathcal{J}_i(\varphi_1) \geq 0 \quad \text{and} \quad \mathcal{J}_i(\varphi_2) \geq 0.$$

Moreover, according to the linearity of $\mathcal{J}_i$, we have that

$$\frac{m}{2}\mathcal{J}_i(\varphi_0) \leq \mathcal{J}_i(\varphi) \leq \frac{M}{2}\mathcal{J}_i(\varphi_0).$$

If $\mathcal{J}_i(\varphi_0) = 0$, then the statement of Theorem 4.9 holds trivially. If $\mathcal{J}_i(\varphi_0) \neq 0$, then

$$m \leq 2\frac{\mathcal{J}_i(\varphi)}{\mathcal{J}_i(\varphi_0)} \leq M.$$

Since φ'' is continuous and $\text{Im}(\varphi'') = [m, M]$, by the intermediate value theorem, there exists $c_i \in [a, b]$ such that

$$\varphi''(c_i) = 2\frac{\mathcal{J}_i(\varphi)}{\mathcal{J}_i(\varphi_0)}.$$

The proof is complete. □

Theorem 4.10 *Let $i \in \{1, 2\}$. If $\varphi, \psi \in C^2([a, b])$, $\mathcal{J}_i(\psi) \neq 0$, then there exists $c_i \in [a, b]$ such that*

$$\frac{\mathcal{J}_i(\varphi)}{\mathcal{J}_i(\psi)} = \frac{\varphi''(c_i)}{\psi''(c_i)}.$$

Proof Consider the function h defined as follows:

$$h(x) = \mathcal{J}_i(\varphi)\psi(x) - \mathcal{J}_i(\psi)\varphi(x).$$

Since $h \in C^2([a, b])$, then by Theorem 4.9 there exists $c_i \in [a, b]$ such that

$$\mathcal{J}_i(h) = \frac{1}{2}h''(c_i)\mathcal{J}_i(\varphi_0).$$

From the definition of h, we get $\mathcal{J}_i(h) = 0$ and we can conclude that $h''(c_i)\mathcal{J}_i(\varphi_0) = 0$. The number $\mathcal{J}_i(\varphi_0)$ is not equal to 0 because in that case $\mathcal{J}_i(\psi) = 0$, and this is a contradiction with the assumption in the theorem. Hence, $h''(c_i) = 0$. Since $h''(c_i) = \mathcal{J}_i(\varphi)\psi''(c_i) - \mathcal{J}_i(\psi)\varphi''(c_i)$, we get the statement, so the proof is complete. □

Remark 4.1 Concerning the "Jensen-gap" functional

$$\mathcal{J}(\varphi, \mu, f) = \int_\Omega \varphi(f(s))\,d\mu(s) - \varphi\left(\int_\Omega f(s)\,d\mu(s)\right)$$

some new information and applications (including refinements) can be found in the paper [1]. For the more general functional $\mathcal{J}_1$ related to our continuous case, a lot of interesting results are also given in the monograph [60].

4.4 Properties of Some Gauss-Pólya Functionals

There are several functionals which are connected with inequalities from Sect. 1.4. These functionals have properties which lead to refinements and improvements of Gauss-Pólya-type inequalities (see [92]).

Let the functions g and p satisfy the assumptions of Theorem 1.11 and define the following functional:

$$G(f) = \exp\left[\int_X \log\left(\int_a^b g'_t(x,t) f(t)\,dt\right) \frac{d\mu(x)}{p(x)}\right]$$
$$- \int_a^b \left[\exp\left(\int_X \log g(x,t) \frac{d\mu(x)}{p(x)}\right)\right]' f(t)\,dt.$$

Under the assumptions of Theorem 1.11(i) G is nonpositive, while under the assumptions of Theorem 1.11 (ii) G is nonnegative. Also, G is positively homogeneous. The property of superadditivity is obtained in the following theorem:

Theorem 4.11 *Let the functions p and g satisfy the assumptions of Theorem 1.11. Let f_1 and f_2 be nonnegative monotone in the same sense such that $G(f_1)$ and $G(f_2)$ are well-defined. Then*

$$G(f_1 + f_2) \geq G(f_1) + G(f_2).$$

Moreover, if $g(x,a) = 0$ for all $x \in X$ and if f_1 and f_2 are nonnegative and nonincreasing functions such that $f_2 \geq f_1$, $f_2 - f_1$ is nonincreasing, and $G(f_1)$, $G(f_2)$ and $G(f_2 - f_1)$

are well-defined, then

$$G(f_2) \geq G(f_1).$$

Proof Let us estimate the difference $G(f_1 + f_2) - G(f_1) - G(f_2)$:

$$\begin{aligned}
G(f_1 + f_2) &- G(f_1) - G(f_2) \\
&= \exp\left[\int_X \log\left(\int_a^b g_t'(x,t)[f_1(t) + f_2(t)]\,dt\right) \frac{d\mu(x)}{p(x)}\right] \\
&\quad - \exp\left[\int_X \log\left(\int_a^b g_t'(x,t) f_1(t)\,dt\right) \frac{d\mu(x)}{p(x)}\right] \\
&\quad - \exp\left[\int_X \log\left(\int_a^b g_t'(x,t) f_2(t)\,dt\right) \frac{d\mu(x)}{p(x)}\right] \\
&\quad - \int_a^b \left[\exp\left(\int_X \log g(x,t) \frac{d\mu(x)}{p(x)}\right)\right]' [f_1(t) + f_2(t)]\,dt \\
&\quad + \int_a^b \left[\exp\left(\int_X \log g(x,t) \frac{d\mu(x)}{p(x)}\right)\right]' f_1(t)\,dt \\
&\quad + \int_a^b \left[\exp\left(\int_X \log g(x,t) \frac{d\mu(x)}{p(x)}\right)\right]' f_2(t)\,dt \\
&= \exp\left[\int_X \log\left(\int_a^b g_t'(x,t)[f_1(t) + f_2(t)]\,dt\right) \frac{d\mu(x)}{p(x)}\right] \\
&\quad - \exp\left[\int_X \log\left(\int_a^b g_t'(x,t) f_1(t)\,dt\right) \frac{d\mu(x)}{p(x)}\right] \\
&\quad - \exp\left[\int_X \log\left(\int_a^b g_t'(x,t) f_2(t)\,dt\right) \frac{d\mu(x)}{p(x)}\right] \geq 0.
\end{aligned}$$

Here we used Corollary 1.1 with $m = 2$, $w_i = 1$, $a_i(x) = [g_t'(x,t) f_i(t)]^{\frac{1}{p(x)}}$, $i = 1, 2$.
Since $G(f_2 - f_1) \geq 0$, we have

$$G(f_2) = G(f_1 + (f_2 - f_1)) \geq G(f_1) + G(f_2 - f_1) \geq G(f_1).$$

The proof is complete. □

4.4 Properties of Some Gauss-Pólya Functionals

Having in mind Theorem 1.12 we define the functional GP as

$$GP(f) = \exp\left(\int_X \log[a(x)p(x)+1]\frac{dx}{p(x)}\right) \times$$

$$\times \exp\left\{\int_X \left[\log\left(\int_a^b f(t)g'(t)g(t)^{a(x)p(x)}\,dt\right)\right]\frac{dx}{p(x)}\right\}$$

$$-\left(1+\int_X a(x)\,dx\right)\int_a^b g(t)^{\int_X a(x)\,dx} g'(t)f(t)\,dt.$$

Using the same method of proof we have results about positivity, superadditivity, and monotonicity, which generalize Theorem 2.1 from [124].

Moreover, we can define the new functional G_W as follows:

$$G_W(f) = \exp\left[\int_X \log\left(\frac{\int_a^b w_x(t)f(t)\,dt}{\int_a^b w_x(t)\,dt}\right)\frac{1}{p(x)}\,dx\right] - \frac{\int_a^b w(t)f(t)\,dt}{\int_a^b w(t)\,dt}$$

and under the conditions in Theorem 1.13 we have that the functional G_W is nonnegative and superadditive. More details about the above-mentioned and other functionals can be found in [92].

While the previous functionals had a monotonic function f as a variable, the following functionals will depend on the weights. Let w be a positive function on X and let denote $W = \int_X w(x)d\mu(x)$. The functionals G_1 and G_2 are defined as follows:

$$G_1(w) = \left\{\int_a^b \left[\exp\left(\int_X \frac{w(x)}{W}\log g(x,t)\,d\mu(x)\right)\right]' f(t)\,dt\right\}^W,$$

$$G_2(w) = \frac{\exp\left[\int_X \log\left(\int_a^b g'_t(x,t)f(t)\,dt\right)w(x)\,d\mu(x)\right]}{G_1(w)}.$$

Theorem 4.12 *Let functions f and g satisfy the assumptions from Theorem 1.11(ii). Let r and s be positive functions on X such that G_1 and G_2 are well-defined for them. Then*

$$G_1(r+s) \leq G_1(r) \cdot G_1(s), \tag{4.8}$$

$$G_2(r+s) \geq G_2(r) \cdot G_2(s). \tag{4.9}$$

If, additionally, $r \geq s$ such that $G_2(r-s)$ is well-defined, then

$$G_2(r) \geq G_2(s).$$

Proof We use the following abbreviations:

$$R = \int_X r(x)\,d\mu(x) \quad \text{and} \quad S = \int_X s(x)\,d\mu(x).$$

Inequality (1.34) for $n = 2$ and nonincreasing f on $[a, b]$ collapses to the following form:

$$\int_a^b \left(x_1^{\frac{1}{p_1}}(t) x_2^{\frac{1}{p_2}}(t)\right)' f(t)\,dt \le \left[\int_a^b x_1'(t) f(t)\,dt\right]^{\frac{1}{p_1}} \left[\int_a^b x_2'(t) f(t)\,dt\right]^{\frac{1}{p_2}}. \quad (4.10)$$

Putting in (4.10):

$$p_1 = \frac{R+S}{R}, \quad x_1(t) = \exp\left(\int_X \frac{r(x)}{R} \log g(x,t)\,d\mu(x)\right),$$

$$p_2 = \frac{R+S}{S}, \quad x_2(t) = \exp\left(\int_X \frac{s(x)}{S} \log g(x,t)\,d\mu(x)\right),$$

we get

$$\int_a^b \left\{\left[\exp\left(\int_X \frac{r(x)}{R} \log g(x,t)\,d\mu(x)\right)\right]^{\frac{R}{R+S}} \right.$$

$$\left. \cdot \left[\exp\left(\int_X \frac{s(x)}{S} \log g(x,t)\,d\mu(x)\right)\right]^{\frac{S}{R+S}} \right\}' f(t)\,dt$$

$$\le \left[\int_a^b \left(\exp\left(\int_X \frac{r(x)}{R} \log g(x,t)\,d\mu(x)\right)\right)' f(t)\,dt\right]^{\frac{R}{R+S}}$$

$$\cdot \left[\int_a^b \left(\exp\left(\int_X \frac{s(x)}{S} \log g(x,t)\,d\mu(x)\right)\right)' f(t)\,dt\right]^{\frac{S}{R+S}}.$$

In other words,

$$\int_a^b \left\{\exp\left[\int_X \frac{r(x)}{R+S} \log g(x,t)\,d\mu(x) + \right.\right.$$

$$\left.\left. + \int_X \frac{s(x)}{R+S} \log g(x,t)\,d\mu(x)\right]\right\}' f(t)\,dt$$

$$\le \left\{\left[\int_a^b \left(\exp\left(\int_X \frac{r(x)}{R} \log g(x,t)\,d\mu(x)\right)\right)' f(t)\,dt\right]^R\right..$$

4.4 Properties of Some Gauss-Pólya Functionals

$$\cdot \left[\int_a^b \left(\exp \left(\int_X \frac{s(x)}{S} \log g(x,t) \, d\mu(x) \right) \right)' f(t) \, dt \right]^S \bigg\}^{\frac{1}{R+S}}.$$

Using the definition of G_1, we have

$$\int_a^b \left\{ \exp \left[\int_X \frac{r(x)+s(x)}{R+S} \log g(x,t) \, d\mu(x) \right] \right\}' f(t) \, dt$$
$$\leq (G_1(r) \cdot G_1(s))^{\frac{1}{R+S}}.$$

Powering by $R+S$, we get $G_1(r+s) \leq G_1(r) \cdot G_1(s)$. The above inequality is used in the proof of the property for G_2. We have

$$G_2(r+s) = \frac{\exp \left[\int_X \log \left(\int_a^b g_t'(x,t) f(t) \, dt \right) (r(x)+s(x) \, d\mu(x)) \right]}{G_1(r+s)}$$

$$\geq \frac{1}{G_1(r)G_1(s)} \exp \left[\int_X \log \left(\int_a^b g_t'(x,t) f(t) \, dt \right) r(x) \, d\mu(x) \right] \cdot$$

$$\cdot \exp \left[\int_X \log \left(\int_a^b g_t'(x,t) f(t) \, dt \right) s(x) \, d\mu(x) \right]$$

$$= G_2(r) G_2(s).$$

By Theorem 1.11(ii) we get that $G_2(r-s) \geq 1$. Using the above inequality, we have

$$G_2(r) = G_2((r-s)+s) \geq G_2(r-s) G_2(s) \geq G_2(s).$$

The proof is complete. □

Remark 4.2 If r and s are weights with the property that there exist positive numbers m and M such that

$$mr \leq s \leq Mr,$$

and if the assumptions of the previous theorem hold, then

$$G_2^m(r) \leq G_2(s) \leq G_2^M(r). \tag{4.11}$$

As the direct consequence of the above remark, we have the following refinement and improvement of the continuous form of the Gauss-Pólya inequality.

Corollary 4.3 *Let f and g satisfy the assumptions from Theorem 1.11(ii) and $\mu(X) = \int_X d\mu(x) \neq 0$ and finite. Let w be a positive function on X such that G_1 and G_2 are well-defined for w and for the constant function. Then*

$$\left\{ \frac{\exp\left[\int_X \log\left(\int_a^b g'_t(x,t) f(t)\, dt\right) d\mu(x)\right]}{\left(\int_a^b \left[\exp\left(\int_X \log g(x,t) \frac{d\mu(x)}{\mu(X)}\right)\right]' f(t)\, dt\right)^{\mu(X)}} \right\}^{w_{\min}}$$

$$\leq \frac{\exp\left[\int_X \log\left(\int_a^b g'_t(x,t) f(t)\, dt\right) w(x) d\mu(x)\right]}{\left(\int_a^b \left[\exp\left(\int_X \log g(x,t) \frac{w(x)}{W} d\mu(x)\right)\right]' f(t)\, dt\right)^{W}}$$

$$\leq \left\{ \frac{\exp\left[\int_X \log\left(\int_a^b g'_t(x,t) f(t)\, dt\right) d\mu(x)\right]}{\left(\int_a^b \left[\exp\left(\int_X \log g(x,t) \frac{d\mu(x)}{\mu(X)}\right)\right]' f(t)\, dt\right)^{\mu(X)}} \right\}^{w_{\max}},$$

where $w_{\min} = \min\{w(x) : x \in X\}$ and $w_{\max} = \max\{w(x) : x \in X\}$.

Proof It is a consequence of (4.11) for weights $w_{\min}$, w, and $w_{\max}$. □

Some Classical Inequalities Involving Banach Lattice Norms

The main focus of this chapter is to describe shortly our present knowledge concerning some classical inequalities (including those by Hölder, Minkowski, Popoviciu, Bellman, Beckenbach-Dresher, and Hardy) in Banach lattice norm settings. We hope that this knowledge can inspire researchers to prove similar versions of other classical inequalities and find new applications outside what we have pointed out in the theory of interpolation between infinitely many Banach spaces (see Appendix A) and in the theory of Hardy-type inequalities.

But, first we present some preliminaries.

Let (Y, ν) be a σ-finite measure space and let $L_0(Y)$ denote the space of ν-measurable functions defined and being finite a.e. on Y. A Banach subspace $(E, \|.\|)$ of $L_0(Y)$ is a Banach lattice (Banach function space) on (Y, ν) if for every $f \in E$, $g \in L_0(Y)$, $|g| \leq |f|$, ν—a.e., it follows that $g \in E$ and $\|g\| \leq \|f\|$.

Moreover, the "convexification" of E, denoted by E^p, $-\infty < p < \infty$, $p \neq 0$, consists of all $f \in L_0(Y)$, satisfying

$$\|f\|_{E^p} = \left(\||f|^p\|_E\right)^{\frac{1}{p}} < \infty.$$

For the case $p < 0$, we assume that $f(t) \neq 0$ for all $t \in Y$. Some classical inequalities are known to hold also in the frame of such Banach lattice norms. See for example [111] and [112]. Those results are presented in the following sections.

It is also known that some classical inequalities for finite many functions (like those of Hölder's and Minkowski's) in L_p and l_p spaces can be generalized to hold for continuous (infinitely) many functions. For such results we refer to the previous chapters, to article [97] and the references there.

But there exists a generalization of the Hölder inequality in both of these directions simultaneously, see our Theorem 5.2.

5.1 Hölder- and Minkowski-Type Inequalities

It is known that if $\|.\|_E$ is a Banach function norm, then $\|f(x,.)\|_E$ need not to be a measurable function. But it is also known that if E has the Fatou property, then indeed $\|f(x,.)\|_E$ is measurable (see [76]). Therefore, for simplicity, we assume that the considered Banach lattices have the Fatou property. It is also known that in this situation E is a perfect space, i.e., $E = E''$, where E'' denotes the second associate space of E.

Our first result is connected to the Hölder inequality. We present the symmetric version of the Hölder inequality, which was published in [112].

Theorem 5.1 *Let p, q, r be nonzero real numbers satisfying $\frac{1}{p} + \frac{1}{q} = \frac{1}{r}$.*

(i) *If $p > 0, q > 0$ and $r > 0$ or if $p < 0, q > 0$ and $r < 0$ or if $p > 0, q < 0$ and $r < 0$, then*

$$\|fg\|_{E^r} \leq \|f\|_{E^p} \|g\|_{E^q}. \tag{5.1}$$

(ii) *If $p > 0, q < 0$ and $r > 0$ or if $p < 0, q > 0$ and $r > 0$ or if $p < 0, q < 0$ and $r < 0$, then*

$$\|fg\|_{E^r} \geq \|f\|_{E^p} \|g\|_{E^q}. \tag{5.2}$$

Proof Let us suppose that $p > 0, q > 0$ and $r > 0$. We put $\bar{f} = \dfrac{f}{\|f\|_{E^p}}$ and $\bar{g} = \dfrac{g}{\|g\|_{E^q}}$ and $\Omega_0 = \{t \in X : \bar{f}\bar{g}(t) \neq 0\}$. Since function $u \mapsto \exp(u)$ is a convex function, we have

$$|\bar{f}(t)|^r |\bar{g}(t)|^r = \exp\left(r \log(|\bar{f}(t)||\bar{g}(t)|)\right)$$

$$= \exp\left(\frac{r}{p} \log(|\bar{f}(t)|^p) + \frac{r}{q} \log(|\bar{g}(t)|^q)\right) \leq \frac{r}{p}|\bar{f}(t)|^p + \frac{r}{q}|\bar{g}(t)|^q.$$

Therefore,

$$\|\bar{f}\bar{g}\|_{E^r}^r \leq \frac{r}{p}\|\bar{f}\|_{E^p}^p + \frac{r}{q}\|\bar{g}\|_{E^q}^q = \frac{r}{p} + \frac{r}{q} = 1,$$

and (5.1) follows.

Let $p > 0, q < 0$ and $r > 0$. Replacing in (5.1) p, q, r with $\frac{r}{p}, -\frac{q}{p}$ and 1, respectively, we obtain

$$\|f\|_{E^p}^p = \||fg|^p|g|^{-p}\|_E \leq \||fg|^p\|_{E^{r/p}} \||g|^{-p}\|_{E^{-q/p}} = \|fg\|_{E^r}^p \|g\|_{E^q}^{-p}.$$

5.1 Hölder- and Minkowski-type inequalities

Hence $\|fg\|_{E^r} \geq \|f\|_{E^p}\|g\|_{E^q}$. In the same way we find that this inequality holds for the symmetric case $p < 0, q > 0$, and $r > 0$.

Let $p < 0, q > 0$, and $r < 0$ or $p > 0, q < 0$ and $r < 0$. Replacing in (5.2) p, q, r with $-p, -q$, and $-r$, respectively, we get

$$\|fg\|_{E^r} = \||fg|^{-1}\|_{E^{-r}}^{-1} \leq \||f|^{-1}\|_{E^{-p}}^{-1}\||g|^{-1}\|_{E^{-q}}^{-1} = \|f\|_{E^p}\|g\|_{E^q}.$$

Let $p < 0, q < 0$ and $r < 0$. From (5.1), it follows that

$$\|fg\|_{E^r} = \||fg|^{-1}\|_{E^{-r}}^{-1} \geq \||f|^{-1}\|_{E^{-p}}^{-1}\||g|^{-1}\|_{E^{-q}}^{-1} = \|f\|_{E^p}\|g\|_{E^q}.$$

The proof is complete. □

Remark 5.1 As an immediate consequence of Theorem 5.1, we get that if $p, q > 1$, $\frac{1}{p} + \frac{1}{q} = 1$, then

$$\|fg\|_E \leq \|f\|_{E^p}\|g\|_{E^q}. \tag{5.3}$$

If $p < 1, p \neq 0$, then (5.3) holds in the reverse direction. All these inequalities are sharp and we even have equality when $g = cf^{p-1}$, where c is a constant. In particular, if E is $L_1(\Omega)$, where (Ω, μ) is a measure space, (5.3) coincides with the standard Hölder inequality.

It is also worth noting that the corresponding inequalities hold for n functions, ($n = 3, 4, \ldots$) and also here all inequalities are sharp.

We also need the following more general form of the Hölder inequality (both continuous and involving Banach function norms):

Theorem 5.2 *Let $E = E''$, $0 < b \leq \infty$, $p(x) > 0$, $u(x) \geq 0$ be measurable and define p by*

$$\frac{1}{p} = \int_0^b \frac{u(x)}{p(x)} \, dx,$$

where $\int_0^b u(x) \, dx = 1$. Then

$$\left\|\exp\left(\int_0^b \log f(x, y)u(x) \, dx\right)\right\|_{E^p} \leq \exp\left(\int_0^b \log \|f(x, y)\|_{E^{p(x)}} u(x) \, dx\right). \tag{5.4}$$

A proof of this result can be found in [90]. Here we present a slightly modified proof.

Proof Dividing inequality (5.4) with the expression on the right-hand side of (5.4), we get an equivalent inequality:

$$\left\| \exp\left(\int_0^b \log a(x,y) u(x) dx \right) \right\|_{E^p} \leq 1, \tag{5.5}$$

where $a(x,y) = \dfrac{f(x,y)}{\|f(x,y)\|_{E^{p(x)}}}$. It is clear that $\|(a(x,y))^{p(x)}\|_E = 1$.

Since the function $\exp$ is convex, using the Jensen inequality for $(a(x,y))^{p(x)}$ and after some simple transformations, we get

$$\left(\exp\left(\int_0^b \log a(x,y) u(x) dx \right) \right)^p = \exp\left(p \int_0^b \log a(x,y) u(x) dx \right)$$

$$= \exp\left(\int_0^b \log(a(x,y))^{p(x)} \frac{pu(x)}{p(x)} dx \right)$$

$$\leq \frac{1}{\int_0^b \frac{pu(x)}{p(x)} dx} \int_0^b (a(x,y))^{p(x)} \frac{pu(x)}{p(x)} dx$$

$$= \int_0^b (a(x,y))^{p(x)} \frac{pu(x)}{p(x)} dx. \tag{5.6}$$

Let us denote $F(x,y) = (a(x,y))^{p(x)} \frac{pu(x)}{p(x)}$. It is easy to see that $\|F(x,y)\|_E = \frac{pu(x)}{p(x)}$. Since $E = E''$, using the variant of the Minkowski inequality given in [59, Chapter 2], we get

$$\left\| \int_0^b F(x,y) dx \right\|_E \leq \int_0^b \|F(x,y)\|_E dx = \int_0^b \frac{pu(x)}{p(x)} dx = 1. \tag{5.7}$$

According to (5.6) and (5.7), we conclude that (5.5) and, thus, (5.4) holds and the proof is complete. □

Example 5.1 Let $b = 1$, $n = 2, 3, \ldots$, $p_1, p_2, \ldots, p_n > 0$, with $\sum_{i=1}^n \dfrac{1}{p_i} = \dfrac{1}{p}$, $u(x) = p(x) = 1$ on $[0, 1]$ and $f(x,y) = f_1(y)$ on $0 \leq x < \dfrac{1}{p_1}$, $f(x,y) = f_2(y)$ on $\dfrac{1}{p_1} \leq x < \dfrac{1}{p_1} + \dfrac{1}{p_2}$, $\ldots$, $f(x,y) = f_n(y)$ on $\sum_{i=1}^{n-1} \dfrac{1}{p_i} \leq x \leq \dfrac{1}{p}$. Then (5.4) reads

$$\left\| \prod_{i=1}^n f_i^{1/p_i} \right\|_{E^p} \leq \prod_{i=1}^n \|f_i^{1/p_i}\|_{E^{p_i}}.$$

5.1 Hölder- and Minkowski-type inequalities

In particular, if $E = L_1(X)$, where (X, μ) is a measure space, this is just slightly rewritten form of the standard Hölder inequality (with n functions involved) as an inequality between standard geometric means.

Corollary 5.1 *Let u and v be weight functions on $[0, b]$, $0 < b \leq \infty$, and on the measure space Y, respectively, such that $\int_0^b u(x)\,dx = 1$. Let $p(x) > 0$ be measurable and define p by*

$$\frac{1}{p} = \int_0^b \frac{u(x)}{p(x)}\,dx.$$

Then

$$\int_Y \exp\left(\int_0^b \log f(x, y) u(x)\,dx\right) v(y)\,dy$$
$$\leq \exp\left(\int_0^b \log\left(\int_Y f(x, y)^{\frac{p(x)}{p}} v(y)\,dy\right) \frac{u(x)p}{p(x)}\,dx\right). \qquad (5.8)$$

Remark 5.2 In particular, if we take $p(x) = p$, then (5.8) coincides with (1.4) in Theorem 1.1 with $X = [0, b]$ and $d\nu(y) = dy$.

Next we formulate a somewhat more general result which was proved in [90], but first we need some more definitions.

If A is a Banach function space on Ω and $w = w(s), s \in \Omega$ is a positive weight on Ω, then the space $E^p(A, w), 0 < p < \infty$, consists of all strongly measurable functions $x = x(s)$ satisfying

$$\|x\|_{E^p(A,w)} = \left[\left\|(\|x(s)\|_A w(s))^p\right\|_E\right]^{\frac{1}{p}} < \infty.$$

Note that $E^p(\mathbb{R}, 1)$ is the usual E^p space.

Let $\{w_x(s), x \in [0, b), s \in \Omega\}$ denotes a family of weight functions such that $w_x(s)$ are measurable on $[0, b) \times \Omega$ and consider the generalized geometric mean

$$w = w(s) = \exp\left(\int_0^b \log w_x(s) u(x)\,dx\right).$$

A more general version of Theorem 5.2 reads:

Theorem 5.3 *Let $E = E''$, $0 < b \leq \infty$, $p(x) > 0$, $u(x) \geq 0$ be measurable and define p by*

$$\frac{1}{p} = \int_0^b \frac{u(x)}{p(x)} \, dx,$$

where $\int_0^b u(x) \, dx = 1$. Let $Z_x = E^{p(x)}(A, w_x)$. Assume that $f_x(s) \in Z_x$, and assume that $u(x) \log \|f_x(s)\|_{Z_x}$ and $u(x) \log \|f_x(s)\|_A$, $x \in \Omega$ are measurable over $[0, b)$. Then

$$\left\| \exp \left(\int_0^b \log f_x(s) u(x) \, dx \right) \right\|_{E^p(w)} \leq \exp \left(\int_0^b \log \|f_x(s)\|_{Z_x} u(x) \, dx \right). \tag{5.9}$$

Proof The proof is similar to that of Theorem 5.2. For details see [90]. □

We can get a variant of Corollary 5.1 if in (5.9) we take $E = L_1$ (and then replace f by $f^{1/p}$).

Remark 5.3 Variants of (5.9) were very guiding for us when we were involved to develop parts of the theory of interpolation between families of Banach spaces (especially complex and Calderón methods), see for example, [31, 57, 58, 88, 89, 91]. In Appendix we describe shortly these ideas that are closely related to continuous forms of inequalities and corresponding generalized forms of inequalities.

Remark 5.4 We note that all versions (5.4), (5.8), and (5.9) are just continuous Hölder-type inequalities written as inequalities between generalized geometric means. This is just a natural generalization of the fact that the classical Hölder inequalities can be written as the corresponding inequalities between classical geometric means.

Let us also recall the following limit reversed Minkowski-type inequality, with geometric means involved. It follows from Corollary 1.1 (see Remark 1.3).

Lemma 5.1 *Let f, g, u be positive measurable functions on X such that $\int_X u(x) d\mu(x) = 1$. Then*

$$\exp \left[\int_X \log f(x) u(x) \, d\mu(x) \right] + \exp \left[\int_X \log g(x) u(x) \, d\mu(x) \right]$$

$$\leq \exp \left[\int_X \log(f(x) + g(x)) u(x) \, d\mu(x) \right].$$

In the rest of this chapter, we will consider different variants of Minkowski's inequality. First, we state and prove the following Minkowski inequality for the space E^p (see [112]).

5.1 Hölder- and Minkowski-type inequalities

Theorem 5.4 *Let $f_i, i = 1, \ldots, n$ be nonnegative measurable functions on the measure space Y.*

(a) If $p \geq 1$, then

$$\left\| \sum_{i=1}^n f_i \right\|_{E^p} \leq \sum_{i=1}^n \|f_i\|_{E^p}. \tag{5.10}$$

(b) If $p < 1$, $p \neq 0$, and $\|\sum_{i=1}^n f_i\|_E \geq \sum_{i=1}^n \|f_i\|_E$, then

$$\left\| \sum_{i=1}^n f_i \right\|_{E^p} \geq \sum_{i=1}^n \|f_i\|_{E^p}.$$

Proof Let $p > 1$. According to Remark 5.1, we have

$$\|f\|_{E^p} = \sup_z \|xz\|_E,$$

where supremum is taken over all nonnegative functions z such that $\|z\|_{E^q} \leq 1$, $q = \frac{p}{p-1}$. Therefore

$$\left\| \sum_{i=1}^n f_i \right\|_{E^p} = \sup_z \left\| \sum_{i=1}^n z f_i \right\|_E \leq \sup_z \sum_{i=1}^n \|z f_i\|_E$$

$$\leq \sum_{i=1}^n \sup_z \|z f_i\|_E = \sum_{i=1}^n \|f_i\|_{E^p}.$$

The case $p = 1$ is trivial, so (5.10) holds for all $p \geq 1$ and the proof of (a) is complete.

By using Theorem 5.1 (b) and Remark 5.1, the proof of (b) can be carried out in a similar way and the proof is complete. □

The following theorem gives a continuous version of the Minkowski inequality involving Banach lattice norms (see [98]).

Theorem 5.5 *Let $E = E''$, $p \geq 1$, $f(x, y) \geq 0$ on $X \times Y$, where X and Y are measure spaces, and let for almost all $x \in X$, $y \mapsto f(x, y) \in E^p$. If the function $\|f^p(x, y)\|_E^{1/p}$ is integrable on X, then*

$$\left\| \int_X f(x, y) \, dx \right\|_{E^p} \leq \int_X \|f(x, y)\|_{E^p} \, dx$$

or, equivalently,

$$\left\| \left(\int_X f(x, y) \, dx \right)^p \right\|_E^{\frac{1}{p}} \leq \int_X \| f^p(x, y) \|_E^{\frac{1}{p}} \, dx. \tag{5.11}$$

Proof The proof follows from the fact that since $E = E''$, i.e., E has the Fatou property, then also E^p, $p \geq 1$, has this property. This means that we can use the Minkowski inequality for perfect spaces, which is proved in [59, Chapter 2]. Note that in the case $E = L_1(Y)$ this is just the classical integral Minkowski inequality. □

Let us describe a p-convexity and q-concavity of Banach lattices.

Definition 5.1 The Banach lattice E is p-convex or q-concave if there exists a positive constant M such that, for every finite set $x_1, x_2, \ldots, x_n$ of elements in E we have

$$\left\| \left(\sum_{i=1}^n |x_i|^p \right)^{1/p} \right\|_E \leq M \left(\sum_{i=1}^n \|x_i\|_E^p \right)^{1/p},$$

or

$$\left(\sum_{i=1}^n \|x_i\|_E^q \right)^{1/q} \leq M \left\| \left(\sum_{i=1}^n |x_i|^q \right)^{1/q} \right\|_E,$$

respectively. The smallest constant M satisfying the corresponding inequality is called the constant of p-convexity, respectively, of q-concavity.

In [119] the following fact was proved: Let ρ and λ be function norms with the Fatou property and assume that there exists p, $1 \leq p \leq \infty$, such that ρ is p-convex and λ is p-concave. Then there exists a constant C such that, for all measurable $f(x, y)$, we have

$$\rho(\lambda(f(x, .))) \leq C\lambda(\rho(f(., y))). \tag{5.12}$$

If we take $\rho(h)$ to be $\int_X h(x) \, dx$ which is 1-convex with constant of convexity equal to 1 and $\lambda = \| \ \|_E$, where E is 1-concave with constant of concavity equal to M, and follow the part of the proof of (5.12) in which the constant of convexity of the norm $\rho(.)$ is equal to 1, we find that

$$M \left\| \int_X f(x, y) \, dx \right\|_E \geq \int_X \|f(x, y)\|_E \, dx. \tag{5.13}$$

Next we state and prove a related result for the E^p spaces (see [98]).

Theorem 5.6 *If E has the Fatou property and is 1-concave with constant of concavity equal to M, $p < 1$, $p \neq 0$, then*

$$M \left\| \int_X f(x,y)\,dx \right\|_{E^p} \geq \int_X \|f(x,y)\|_{E^p}\,dx. \tag{5.14}$$

Proof As usual, denote $q = \frac{p}{p-1}$. Using (5.13) and the sharpness of Hölder inequalities for E^p spaces (see Remark 5.1), we have

$$M \left\| \int_X f(x,y)\,dx \right\|_{E^p} = \inf_{\|z\|_{E^q}=1} M \left\| \int_X f(x,y)z\,dx \right\|_E$$

$$\geq \inf_{\|z\|_{E^q}=1} \int_X \|f(x,y)z\|_E\,dx \geq \int_X \inf_{\|z\|_{E^q}=1} \|f(x,y)z\|_E\,dx$$

$$= \int_X \|f(x,y)\|_{E^p}\,dx.$$

The proof is complete. □

5.2 Popoviciu- and Bellman-Type Inequalities

In this section we present our results concerning reverse versions of the Hölder and Minkowski inequalities involving Banach lattice norms (see [98]). Following the same terminology we used in the classical case, we call them the Popoviciu- and Bellman-type inequalities. Our first main result reads:

Theorem 5.7 *Let E be a Banach function space, such that $E'' = E$ and $0 < b \leq \infty$. Let $p(x)$ be positive and measurable, $u(x)$ be nonnegative and measurable, and define p by*

$$\frac{1}{p} = \int_0^b \frac{u(x)}{p(x)}\,dx,$$

where $\int_0^b u(x)\,dx = 1$. If $f(x,y)$ is positive and $f_0(x) > \|f(x,y)\|_{E^{p(x)}} > 0$, then

$$\exp\left(\int_0^b \log f_0(x) u(x)\,dx\right) - \left\|\exp\left(\int_0^b \log f(x,y) u(x)\,dx\right)\right\|_{E^p}$$

$$\geq \exp\left[\int_0^b \log\left(f_0(x) - \|f(x,y)\|_{E^{p(x)}}\right) u(x)\,dx\right], \tag{5.15}$$

provided that all integrals which occur in (5.15) exist.

Proof By putting in Lemma 5.1

$$X = [0, b), \quad f(x) = f_0(x) - \|f(x, y)\|_{E^{p(x)}}, \quad g(x) = \|f(x, y)\|_{E^{p(x)}},$$

we get

$$\exp\left[\int_0^b \log\left(f_0(x) - \|f(x, y)\|_{E^{p(x)}}\right) u(x)\, dx\right]$$
$$+ \exp\left(\int_0^b \log(\|f(x, y)\|_{E^{p(x)}}) u(x)\, dx\right) \leq \exp\left(\int_0^b \log f_0(x) u(x)\, dx\right).$$

Hence, by using this estimate and Theorem 5.2, we find that

$$\exp\left(\int_0^b \log f_0(x) u(x)\, dx\right) - \left\|\exp\left(\int_0^b \log f(x, y) u(x)\, dx\right)\right\|_{E^p}$$
$$\geq \exp\left(\int_0^b \log f_0(x) u(x)\, dx\right) - \exp\left(\int_0^b \log(\|f(x, y)\|_{E^{p(x)}}) u(x)\, dx\right)$$
$$\geq \exp\left[\int_0^b \log\left(f_0(x) - \|f(x, y)\|_{E^{p(x)}}\right) u(x)\, dx\right].$$

Thus, (5.15) is proved and the proof is complete. □

Example 5.2 Applying Theorem 5.7 with $E = L_1(Y)$, we find that

$$\exp\left(\int_0^b \log f_0(x) u(x)\, dx\right)$$
$$- \left[\int_Y \exp\left(\int_0^b \log f(x, y) u(x)\, dx\right)^p v(y)\, d\nu(y)\right]^{\frac{1}{p}}$$
$$\geq \exp\left[\int_0^b \log\left(f_0(x) - \left(\int_Y f^{p(x)}(x, y) v(y)\, d\nu(y)\right)^{\frac{1}{p(x)}}\right) u(x)\, dx\right],$$

which in the case $p(x) \equiv 1$ becomes the known result (1.23) for $X = [0, b)$ and $v_0 = 1$.

Next we state a result not for infinitely many functions, but just for two of them.

Theorem 5.8 *Let E be a Banach function space, and let $f, g \geq 0$ and $p, q \neq 0$, where $\frac{1}{p} + \frac{1}{q} = 1$.*

5.2 Popoviciu- and Bellman-type inequalities

(a) Let $p \geq 1$. If $c_1^p - \|f^p\|_E \geq 0$, $c_2^q - \|g^q\|_E \geq 0$, then

$$c_1 c_2 - \|fg\|_E \geq \left(c_1^p - \|f^p\|_E\right)^{1/p} \left(c_2^q - \|g^q\|_E\right)^{1/q}. \qquad (5.16)$$

(b) Let $0 < p < 1$. If $\|g^q\|_E > 0$, $c_2^q - \|g^q\|_E > 0$, then the reverse inequality (5.16) holds.

(c) Let $p < 0$. If $\|f^p\| > 0$, $c_1^p - \|f^p\|_E > 0$, then the reverse inequality (5.16) holds.

Proof (a) Let $p > 1$. Let $c_1^p - \|f^p\|_E$, $c_2^q - \|g^q\|_E$ be strictly positive. Let $X_1 \cup X_2 = [0, b)$, $X_1 \cap X_2 = \emptyset$, $\int_{X_1} u(x)\, dx = \frac{1}{p}$, $\int_{X_2} u(x)\, dx = \frac{1}{q}$. We can get the result like a corollary from the inequality in Theorem 5.7, by taking $p(x) = 1$,

$$f_0(x) = \begin{cases} c_1^p, & x \in X_1 \\ c_2^q, & x \in X_2 \end{cases} \qquad f(x, y) = \begin{cases} f^p(y), & x \in X_1 \\ g^q(y), & x \in X_2. \end{cases}$$

If, for instance, $c_1^p - \|f^p\|_E = 0$, $c_2^q - \|g^q\|_E \geq 0$, we have to prove that $c_1 c_2 \geq \|fg\|_E$. For this purpose we just use inequality (5.3), namely

$$\|fg\|_E \leq \|f^p\|_E^{\frac{1}{p}} \|g^q\|_E^{\frac{1}{q}} \leq c_1 c_2.$$

(b) The case $0 < p < 1$ can be treated by using the reverse inequality (5.3), which says that now

$$\|fg\|_E \geq \|f^p\|_E^{\frac{1}{p}} \|g^q\|_E^{\frac{1}{q}}. \qquad (5.17)$$

In this case, the discrete Hölder inequality states that for $a_1, a_2, b_1, b_2 \geq 0$:

$$(a_1^p + a_2^p)^{1/p} (b_1^q + b_2^q)^{1/q} \leq a_1 b_1 + a_2 b_2.$$

Putting in this inequality

$$a_1 = (c_1^p - \|f^p\|_E)^{1/p}, \ a_2 = (\|f^p\|_E)^{1/p}, \ b_1 = (c_2^q - \|g^q\|)\|_E^{1/q}, \ b_2 = (\|g^q\|_E)^{1/q},$$

and using (5.17), we get

$$c_1 c_2 \leq (c_1^p - \|f^p\|_E)^{1/p} (c_2^q - \|g^q\|)\|_E^{1/q} + (\|f^p\|_E)^{1/p} (\|g^q\|_E)^{1/q}$$

$$\leq (c_1^p - \|f^p\|_E)^{1/p} (c_2^q - \|g^q\|)\|_E^{1/q} + \|fg\|_E,$$

which gives us the wanted reversed inequality (5.16).

(c) The case $p < 0$ can be proved analogously, so we omit the details. The proof is complete. □

Corollary 5.2 *Let E be a Banach function space, and let f, g, $c_1^p - \|f^p\|_{E^r} \geq 0$, $c_2^q - \|g^q\|_{E^s} \geq 0$ where $p, q, r, s > 0$, $\frac{1}{p} + \frac{1}{q} = 1$, and $\frac{1}{pr} + \frac{1}{qs} = 1$. Then*

$$c_1 c_2 - \|fg\|_E \geq (c_1^p - \|f^p\|_{E^r})^{1/p} (c_2^q - \|g^q\|_{E^s})^{1/q}. \tag{5.18}$$

Proof When $c_1^p - \|f^p\|_E$, $g_0^q - \|g^q\|_E$ are strictly positive, we can get (5.18) like a corollary from Theorem 5.7, by taking $f_0(x) = c_1^p$ for $x \in X_1$, $f_0(x) = c_2^q$ for $x \in X_2$, $f(x, y) = f^p(y)$, $x \in X_1$, $f(x, y) = g^q(y)$, $x \in X_2$, where $X_1 \cup X_2 = [0, b)$, $X_1 \cap X_2 = \emptyset$, $\int_{X_1} u(x)\, dx = \frac{1}{p}$, $\int_{X_2} u(x)\, dx = \frac{1}{q}$, $p(x) = r$ for $x \in X_1$, $p(x) = s$ for $x \in X_2$.

Look for instance on the third expression in inequality (5.15):

$$\exp\left[\int_0^b \log\left(f_0(x) - \|f(x, y)\|_{E^{p(x)}}\right) u(x)\, dx\right]$$

$$= \exp\left[\int_{X_1} \log\left(c_1^p - \|f^p(y)\|_{E^r}\right) u(x)\, dx\right.$$

$$\left. + \int_{X_2} \log\left(c_2^q - \|g^q(y)\|_{E^s}\right) u(x)\, dx\right]$$

$$= \exp\left[\log\left(c_1^p - \|f^p\|_{E^r}\right)\frac{1}{p} + \log\left(c_2^q - \|g^q\|_{E^s}\right)\frac{1}{q}\right]$$

$$= (c_1^p - \|f^p\|_{E^r})^{1/p} (c_2^q - \|g^q\|_{E^s})^{1/q},$$

which is equal to the right-hand side of inequality (5.18). Similarly we can see that the terms of the left-hand side of (5.15) coincide with the corresponding terms in (5.18).

If, for instance, $c_1^p - \|f^p\|_{E^r} = 0$, $c_2^q - \|g^q\|_{E^s} \geq 0$, we have to prove that $c_1 c_2 \geq \|fg\|$. For this purpose we can just use (5.3) and find that

$$\|fg\|_E \leq \|f^p\|_{E^r}^{\frac{1}{p}} \|g^q\|_{E^s}^{\frac{1}{q}} \leq c_1 c_2.$$

The proof is complete. □

Just as there are results associated with the continuous Hölder inequality, i.e., continuous Popoviciu-type inequalities, so there are results that are related to the continuous Minkowski inequality. We call them continuous Bellman-type inequalities, and they can be seen as the reversed versions of the continuous Minkowski inequality. Our first such Bellman-type inequality reads:

5.2 Popoviciu- and Bellman-type inequalities

Theorem 5.9 *Let X and Y be measure spaces, let $f(x,y)$ be a positive measurable function on $X \times Y$, and assume that $p \geq 1$ and $f_0(x)$ is a function on X such that $f_0^p(x) > \|f^p(x,y)\|_E$, where E is a Banach lattice on Y for all $x \in X$. Assume that E has the Fatou property. Then*

$$\int_X \left(f_0^p(x) - \|f^p(x,y)\|_E\right)^{\frac{1}{p}} dx \leq \left[\left(\int_X f_0(x) dx\right)^p - \left\|\left(\int_X f(x,y) dx\right)^p\right\|_E\right]^{\frac{1}{p}}, \quad (5.19)$$

provided that all integrals exist.

If E is 1-concave with constant of concavity 1, $0 < p < 1$ or $p < 0$ and $\|f^p(x,y)\|_E > 0$, then inequality (5.19) holds in the reverse direction.

Proof Let $p \geq 1$. We consider the following form of the reversed Minkowski integral inequality:

$$\left(\int_X a(x) dx\right)^p + \left(\int_X b(x) dx\right)^p \leq \left[\int_X \left(a^p(x) + b^p(x)\right)^{\frac{1}{p}} dx\right]^p. \quad (5.20)$$

Choosing

$$a(x) = \left(f_0^p(x) - \|f^p(x,y)\|_E\right)^{\frac{1}{p}} \quad \text{and} \quad b(x) = \left(\|f^p(x,y)\|_E\right)^{\frac{1}{p}},$$

in (5.20), we obtain

$$\left(\int_X \left(f_0^p(x) - \|f^p(x,y)\|_E\right)^{\frac{1}{p}} dx\right)^p + \left(\int_X \left(\|f^p(x,y)\|_E\right)^{\frac{1}{p}} dx\right)^p$$

$$\leq \left[\int_X f_0(x) dx\right]^p =: I_1.$$

Next by using (5.11) to the second term in the above inequality, we find that

$$I_1 \geq \left\|\left(\int_X f(x,y) dx\right)^p\right\|_E + \left(\int_X \left(f_0^p(x) - \|f^p(x,y)\|_E\right)^{\frac{1}{p}} dx\right)^p.$$

If $0 < p < 1$, then (5.20) holds in the reversed direction. By using this inequality and instead of (5.11) using (5.14) for $M = 1$, the proof is step by step the same as that for the case $0 < p < 1$. The proof in the case $p < 0$ is similar, so we omit the details. The proof is complete. $\square$

Remark 5.5 Inequality (5.19) can be written as

$$\left(\int_X [f_0^p(x) - \|f(x,y)\|_{E^p}^p]^{\frac{1}{p}} \, dx\right)^p \leq \left[\int_X f_0(x) \, dx\right]^p - \left\|\int_X f(x,y) \, dx\right\|_{E^p}^p.$$

If $E = L_1(Y)$, then we get the result of the first part of the continuous Bellman inequality for $p > 1$, which was stated and proved in [97].

Finally, we state and prove the following Bellman-type inequality:

Theorem 5.10

(a) *Let E be a Banach function space, let $f, g > 0$, $p \geq 1$, $c_1^p - \|f^p\|_E \geq 0$, and $c_2^p - \|g^p\|_E \geq 0$. Then*

$$\left((c_1^p - \|f^p\|_E)^{1/p} + (c_2^p - \|g^p\|_E)^{1/p}\right)^p \leq (c_1 + c_2)^p - \|(f+g)^p\|_E. \quad (5.21)$$

(b) *If $p < 0$ and $c_1, f > 0$, $c_1^p - \|f^p\|_E > 0$, $c_2 > 0$, $g > 0$, $c_2^p - \|g^p\|_E > 0$, then (5.21) holds. If $0 < p < 1$, then the reverse inequality (5.21) holds.*

Proof (a) Let $p > 1$. Let us recall that for $n = 2$, inequality (5.10) has the following form:

$$\|f+g\|_{E^p} \leq \|f\|_{E^p} + \|g\|_{E^p}. \quad (5.22)$$

Taking the pth power of both sides in (5.22), we get

$$\left(\|f^p\|_E^{1/p} + \|g^p\|_E^{1/p}\right)^p = (\|f\|_{E^p} + \|g\|_{E^p})^p$$
$$\geq (\|f+g\|_{E^p})^p = \|(f+g)^p\|_E. \quad (5.23)$$

Let us recall the discrete Minkowski inequality for nonnegative numbers a_1, a_2, b_1, b_2, and $p > 1$:

$$\left((a_1+b_1)^p + (a_2+b_2)^p\right)^{\frac{1}{p}} \leq (a_1^p + a_2^p)^{\frac{1}{p}} + (b_1^p + b_2^p)^{\frac{1}{p}}. \quad (5.24)$$

Put in it:

$$a_1 = (c_1^p - \|f^p\|_E)^{1/p}, \quad b_1 = (c_2^p - \|g^p\|_E)^{1/p},$$
$$a_2 = (\|f^p\|_E)^{1/p}, \quad b_2 = (\|g^p\|_E)^{1/p}. \quad (5.25)$$

Starting with the pth power of the right-hand side in (5.24) with a_1, a_2, b_1, b_2 defined by (5.25) and using (5.23), we obtain that

$$(c_1 + c_2)^p$$
$$\geq \left((c_1^p - \|f^p\|_E)^{1/p} + (c_2^p - \|g^p\|_E)^{1/p}\right)^p + \left((\|f^p\|_E)^{1/p} + (\|g^p\|_E)^{1/p}\right)^p$$
$$\geq \left((c_1^p - \|f^p\|_E)^{1/p} + (c_2^p - \|g^p\|_E)^{1/p}\right)^p + \|(f+g)^p\|_E,$$

so (5.21) is proved.

(b) If $0 < p < 1$, then all inequalities above hold in the reversed directions and the proof follows by just doing obvious modifications of the proof of a). The proof is complete. $\square$

5.3 Beckenbach-Dresher-Type Inequalities

Our main result concerning the Beckenbach-Dresher inequality including its continuous form is given in the following theorem and can be found in [98] and [99]. The proofs of our results are based on continuous form of the Minkowski inequality and direct and reversed integral Hölder inequality for two functions.

Theorem 5.11 *Let E and F be Banach function spaces with the Fatou property.*

(i) *If $0 < u < 1$, $0 < p, q \leq 1$, and E is 1-concave with constant of concavity equal to M, F is 1-concave with constant of concavity equal to N, then the inequality*

$$\frac{\left\|\int_X f(x, y) \, dx\right\|_{E^p}^u}{\left\|\int_X g(x, y) \, dx\right\|_{F^q}^{u-1}} \geq C \int_X \frac{\|f(x, y)\|_{E^p}^u}{\|g(x, y)\|_{F^q}^{u-1}} \, dx \quad (5.26)$$

holds with $C = M^{-u} N^{u-1}$, provided that all above integrals exist.

(ii) *If $u > 1$, $q \leq 1 \leq p$, $q \neq 0$, F is 1-concave with constant of concavity equal to N, then inequality (5.26) holds in the reverse direction with $C = N^{u-1}$, provided that all above integrals exist.*

(iii) *If $u < 0$, $p \leq 1 \leq q$, $p \neq 0$, E is 1-concave with constant of concavity equal to M, then inequality (5.26) holds in the reverse direction with $C = M^{-u}$, provided that all above integrals exist.*

Proof

(i) Let $0 < u < 1, 0 < p, q \leq 1$. Since E is 1-concave with constant of concavity equal to M and F is 1-concave with constant of concavity equal to N, using Theorem 5.6, we have

$$\left\| \int_X f(x, y)\,dx \right\|_{E^p} \geq M^{-1} \int_X \|f(x, y)\|_{E^p}\,dx$$

and

$$\left\| \int_X g(x, y)\,dx \right\|_{F^q} \geq N^{-1} \int_X \|g(x, y)\|_{F^q}\,dx. \qquad (5.27)$$

Since u is positive and $u - 1$ is negative, we get

$$\frac{\left\| \int_X f(x, y)\,dx \right\|_{E^p}^{u}}{\left\| \int_X g(x, y)\,dx \right\|_{F^q}^{u-1}}$$

$$\geq M^{-u} N^{u-1} \left(\int_X \|f(x, y)\|_{E^p}\,dx \right)^u \left(\int_X \|g(x, y)\|_{F^q}\,dx \right)^{1-u}. \qquad (5.28)$$

Putting in the integral Hölder inequality $\int f_1 f_2 \leq \left(\int f_1^r \right)^{1/r} \left(\int f_2^s \right)^{1/s}$ with conjugate exponents r and s functions $f_1 = \|f(x, y)\|_{E^p}^{u}$ and $f_2 = \|g(x, y)\|_{F^q}^{1-u}$ with conjugate exponents $r = \frac{1}{u}$ and $s = \frac{1}{1-u}$, we obtain that

$$\left(\int_X \|f(x, y)\|_{E^p}\,dx \right)^u \left(\int_X \|g(x, y)\|_{F^q}\,dx \right)^{1-u}$$

$$\geq \int_X \|f(x, y)\|_{E^p}^{u} \|g(x, y)\|_{F^q}^{1-u}\,dx. \qquad (5.29)$$

According to (5.28) and (5.29), we can conclude that inequality (5.26) holds.

(ii) Let $u > 1$, and $q \leq 1 \leq p$. Applying Theorem 5.5 for the space E, the following inequality holds:

$$\left\| \int_X f(x, y)\,dx \right\|_{E^p} \geq \int_X \|f(x, y)\|_{E^p}\,dx.$$

Inequality (5.27) holds for the space F, and since in this case both u and $u - 1$ are positive, we find that

5.3 Beckenbach-Dresher-type inequalities

$$\frac{\left\|\int_X f(x,y)\,dx\right\|_{E^p}^u}{\left\|\int_X g(x,y)\,dx\right\|_{F^q}^{u-1}} \leq N^{u-1}\left(\int_X \|f(x,y)\|_{E^p}\,dx\right)^u\left(\int_X \|g(x,y)\|_{F^q}\,dx\right)^{1-u}$$

$$\leq N^{u-1}\int_X \|f(x,y)\|_{E^p}^u \|g(x,y)\|_{F^q}^{1-u}\,dx$$

$$= N^{u-1}\int_X \frac{\|f(x,y)\|_{E^p}^u}{\|g(x,y)\|_{F^q}^{u-1}}\,dx,$$

where in the last inequality we have used the reverse Hölder inequality for conjugate exponents $\frac{1}{u} \in (0,1)$ and $\frac{1}{1-u} < 0$. Hence, also the proof of (ii) is complete. The proof of (iii) is completely similar as the proof of (ii) so we leave out the details. The proof is complete. □

Example 5.3 Let $E = L_1(Y,\nu)$ and $F = L_1(Y,\lambda)$. Then $M = N = 1$, so that if $0 < u \leq 1$, $p \leq 1$, $q \leq 1$, and $p, q \neq 0$, then inequality (5.26) becomes

$$\frac{\left(\int_Y \left(\int_X f(x,y)d\mu(x)\right)^p d\nu(y)\right)^{\frac{u}{p}}}{\left(\int_Y \left(\int_X g(x,y)d\mu(x)\right)^q d\lambda(y)\right)^{\frac{u-1}{q}}} \geq \int_X \frac{\left(\int_Y f^p(x,y)d\nu(y)\right)^{\frac{u}{p}}}{\left(\int_Y g^q(x,y)d\lambda(y)\right)^{\frac{u-1}{q}}}d\mu(x).$$

If $u \geq 1$ and $q \leq 1 \leq p$ ($q \neq 0$) or $u < 0$ and $p \leq 1 \leq q$ ($p \neq 0$), then the inequality is reversed.

Remark 5.6 This result was first stated and proved in [45]. Moreover, a variant of this inequality in time scales settings can be found in [19]. Note also that for the special case $X = Y = L_1(0,\infty)$ and the involved measures are Lebesgue measures we obtain some classical versions of the Beckenbach-Dresher inequality.

Remark 5.7 By applying Theorem 5.11 to appropriate functions, we get a discrete form of the Beckenbach-Dresher inequality in Banach lattice space setting. Namely, if f and g are step functions of the type

$$f(x,y) = f_i(y), \quad g(x,y) = g_i(y) \quad \text{when } x \in [i-1,i),\ i = 1,2,\ldots,n$$

and if $0 < u < 1$, if E is 1-concave with constant of concavity equal to M, if F is 1-concave with constant of concavity equal to N, then the inequality

$$\frac{\left\|\sum_{i=1}^{n} f_i\right\|_E^u}{\left\|\sum_{i=1}^{n} g_i\right\|_F^{u-1}} \geq M^{-u} N^{1-u} \sum_{i=1}^{n} \frac{\|f_i\|_E^u}{\|g_i\|_F^{u-1}}$$

holds.

The following theorem, stated and proved in [98], can be regarded as the reversed version of the Beckenbach-Dresher inequality in Banach lattice spaces.

Theorem 5.12 *Let X, Y, and Z be measure spaces. Let $f_0(x) > \|f(x, y)\|_{E^p}$ for all $x \in X$, let $g(x, z)$ be a positive measurable function on $X \times Z$, and assume that $g_0(x)$ is a function on X such that $g_0(x) > \|g(x, y)\|_{F^q}$ for all $x \in X$, where E is a Banach function space on Y for all $x \in X$ with the Fatou property and F is a Banach function space on Z for all $x \in X$ with the Fatou property.*

(i) *If $0 < u < 1$, $p \geq 1$ and $q \geq 1$, then the following continuous reverse type version of Beckenbach-Dresher's inequality holds:*

$$\frac{\left(\left[\int_X f_0(x)\,dx\right]^p - \left\|\left[\int_X f(x,y)\,dx\right]^p\right\|_E\right)^{\frac{u}{p}}}{\left(\left[\int_X g_0(x)\,dx\right]^q - \left\|\left[\int_X g(x,y)\,dx\right]^q\right\|_F\right)^{\frac{u-1}{q}}}$$
$$\geq \int_X \frac{\left(f_0^p(x) - \|f^p(x,y)\|_E\right)^{\frac{u}{p}}}{\left(g_0^q(x) - \|g^q(x,y)\|_F\right)^{\frac{u-1}{q}}}\,dx. \quad (5.30)$$

(ii) *If $u \geq 1$, $0 < p \leq 1$ and $q \geq 1$, then the reversed inequality (5.30) holds.*
(iii) *If $u < 0$, $0 < q \leq 1$ and $p \geq 1$, then the reversed inequality (5.30) holds.*

In the cases when $p < 1$, an additional condition on the lattice E is that it should be concave with constant of concavity 1, in the cases when $q < 1$ an additional condition on the lattice F is that it should be concave with constant of concavity 1.

Proof Let $0 < u \leq 1$, $p \geq 1$, $q \geq 1$. By using the continuous Bellman inequality in Banach lattices (Theorem 5.9), we get

$$\left(\int_X \left(f_0^p(x) - \|f^p(x,y)\|_E\right)^{\frac{1}{p}}\,dx\right)^p \leq \left[\int_X f_0(x)\,dx\right]^p - \left\|\left[\int_X f(x,y)\,dx\right]^p\right\|_E$$

5.3 Beckenbach-Dresher-type inequalities

and

$$\left(\int_X (g_0^q(x) - \|g^q(x,y)\|_F)^{\frac{1}{q}} \, dx\right)^q \leq \left[\int_X g_0(x) \, dx\right]^q - \left\|\left[\int_X g(x,y) \, dx\right]^q\right\|_F.$$

Since $\frac{u}{p} > 0$ and $\frac{1-u}{q} > 0$, the following two inequalities hold:

$$\left(\int_X (f_0^p(x) - \|f^p(x,y)\|_E)^{\frac{1}{p}} \, dx\right)^u$$
$$\leq \left(\left[\int_X f_0(x) \, dx\right]^p - \left\|\left[\int_X f(x,y) \, dx\right]^p\right\|_E\right)^{\frac{u}{p}} \quad (5.31)$$

$$\left(\int_X (g_0^q(x) - \|g^q(x,y)\|_F)^{\frac{1}{q}} \, dx\right)^{1-u}$$
$$\leq \left(\left[\int_X g_0(x) \, dx\right]^q - \left\|\left[\int_X g(x,y) \, dx\right]^q\right\|_F\right)^{\frac{1-u}{q}}. \quad (5.32)$$

By now multiplying (5.31) by (5.32) and using the integral Hölder inequality for two functions, we get

$$\frac{\left(\left[\int_X f_0(x) \, dx\right]^p - \left\|\left[\int_X f(x,y) \, dx\right]^p\right\|_E\right)^{\frac{u}{p}}}{\left(\left[\int_X g_0(x) \, dx\right]^q - \left\|\left[\int_X g(x,y) \, dx\right]^q\right\|_F\right)^{\frac{u-1}{q}}}$$
$$\geq \left(\int_X (f_0^p(x) - \|f^p(x,y)\|_E)^{\frac{1}{p}} \, dx\right)^u \left(\int_X (g_0^q(x) - \|g^q(x,y)\|_F)^{\frac{1}{q}} \, dx\right)^{1-u}$$
$$\geq \int_X (f_0^p(x) - \|f^p(x,y)\|_E)^{\frac{u}{p}} (g_0^q(x) - \|g^q(x,y)\|_F)^{\frac{1-u}{q}} \, dx$$
$$= \int_X \frac{(f_0^p(x) - \|f^p(x,y)\|_E)^{\frac{u}{p}}}{(g_0^q(x) - \|g^q(x,y)\|_F)^{\frac{u-1}{q}}} \, dx,$$

i.e., (5.30) holds and thus (i) is proved. The proofs of the cases (ii) and (iii) are completely similar, so we leave out the details. The proof is complete. □

5.4 Some New Hardy-Type Inequalities in Banach Function Spaces

First we pronounce again that a lot of information about Hardy-type inequalities is known, see, e.g., the monographs [56, 67, 69, 105] and the references therein. Some new information not found in these books has been reported on in our Sect. 2.5. In particular, the current status concerning Hardy-type inequalities in other function spaces is described in [69, Chapter 7.6]. Here we shall report on some even newer results concerning Hardy-type inequalities in Banach function spaces.

In [12, Theorem 2] the following Hardy-type inequality was given:

Theorem 5.13 *Let $0 < b \leq \infty$, $-\infty \leq a < c \leq \infty$, let φ be a positive convex function on (a, c), and E be a Banach function space on $[0, b)$. If E has the Fatou property and $a < f(x) < c$, then*

$$\left\| \varphi \left(\frac{1}{x} \int_0^x f(t)\, dt \right) \right\|_E \leq \int_0^b \varphi(f(t)) \left\| \frac{1}{x} \chi_{[t,b)}(x) \right\|_E dt, \tag{5.33}$$

provided that both sides have sense.

Proof The proof follows by first using the Jensen inequality, after that using the lattice property of E and the Fubini theorem. □

Here we present a refinement of inequality (5.33) using superquadracity. Our first main result in this section reads (see [103]):

Theorem 5.14 *Let $0 < b \leq \infty$, $-\infty \leq a < c \leq \infty$, let φ be a positive and superquadratic function on (a, c), and E be a Banach function space on $[0, b)$. If E has the Fatou property and $a < f(x) < c$, then*

$$\left\| \varphi \left(\frac{1}{x} \int_0^x f(t)\, dt \right) \right\|_E$$
$$\leq \int_0^b \varphi(f(t)) \left\| \left(1 - \frac{\varphi\left(|f(t) - \frac{1}{x} \int_0^x f(s)\, ds|\right)}{\varphi(f(t))} \right) \frac{1}{x} \chi_{[t,b)}(x) \right\|_E dt, \tag{5.34}$$

provided that both sides have sense.

Proof Let $D = \{(x, t) : 0 \leq x \leq b, 0 \leq t \leq x\}$. Then

$$\chi_D(x, t) = \chi_{[0,x]}(t) = \chi_{[t,b]}(x). \tag{5.35}$$

5.4 Some new Hardy-type inequalities

By using Theorem 3.5 with $d\mu(x) = \frac{1}{x} dx$, the lattice property of E, Theorem 5.5 for $p = 1$ and (5.35), we find that

$$\left\| \varphi\left(\frac{1}{x}\int_0^x f(t)\, dt\right) \right\|_E \le \left\| \int_0^x \frac{\varphi(f(t)) - \varphi\left(\left|f(t) - \frac{1}{x}\int_0^x f(s)\, ds\right|\right)}{x} dt \right\|_E$$

$$= \left\| \int_0^b \frac{\varphi(f(t)) - \varphi\left(\left|f(t) - \frac{1}{x}\int_0^x f(s)\, ds\right|\right)}{x} \chi_{[0,x]}(t)\, dt \right\|_E$$

$$= \left\| \int_0^b \frac{\varphi(f(t)) - \varphi\left(\left|f(t) - \frac{1}{x}\int_0^x f(s)\, ds\right|\right)}{x} \chi_D(x,t)\, dt \right\|_E$$

$$\le \int_0^b \left\| \frac{\varphi(f(t)) - \varphi\left(\left|f(t) - \frac{1}{x}\int_0^x f(s)\, ds\right|\right)}{x} \chi_D(x,t) \right\|_E dt$$

$$= \int_0^b \left\| \frac{\varphi(f(t)) - \varphi\left(\left|f(t) - \frac{1}{x}\int_0^x f(s)\, ds\right|\right)}{x} \chi_{[t,b]}(x) \right\|_E dt$$

$$= \int_0^b \varphi(f(t)) \left\| \left(1 - \frac{\varphi\left(\left|f(t) - \frac{1}{x}\int_0^x f(s)\, ds\right|\right)}{\varphi(f(t))}\right) \frac{1}{x} \chi_{[t,b]}(x) \right\|_E dt,$$

so that (5.34) holds and the proof is complete. □

Here we just give one example of application of Theorem 5.14 (c.f. [104, Proposition 2.1] and [63, Theorem 2.3]):

Corollary 5.3 *Let $0 < b \le \infty$, $u : (0, b) \to \mathbb{R}$ be a nonnegative weight function such that the function $x \mapsto \frac{u(x)}{x^2}$ is locally integrable on $(0, b)$, and define the weight function v by*

$$v(t) = t \int_t^b \frac{u(x)}{x^2}\, dx, \quad t \in (0, b).$$

If the real-valued function φ is positive and superquadratic on (a, c), $0 \le a < c \le \infty$, then the inequality

$$\int_0^b u(x)\varphi\left(\frac{1}{x}\int_0^x f(t)\,dt\right)\frac{dx}{x} \quad (5.36)$$

$$\leq \int_0^b v(t)\varphi(f(t))\frac{dt}{t} - \int_0^b \int_t^b \varphi\left(\left|f(t) - \frac{1}{x}\int_0^x f(s)\,ds\right|\right)\frac{u(x)}{x^2}\,dx\,dt$$

holds for all f with $a < f(x) < c$, $0 < x \leq b$.

Proof It is known that $E = L_1\left((0,b), \frac{u(x)}{x}\,dx\right)$ satisfies the Fatou property (see, e.g., [17]). Moreover, in this case we have

$$\left\|\left(1 - \frac{\varphi\left(|f(t) - \frac{1}{x}\int_0^x f(s)\,ds|\right)}{\varphi(f(t))}\right)\frac{1}{x}\chi_{[t,b]}(x)\right\|_E$$

$$= \int_t^b \frac{u(x)}{x^2}\,dx - \frac{1}{\varphi(f(t))}\int_t^b \varphi\left(\left|f(t) - \frac{1}{x}\int_0^x f(s)\,ds\right|\right)\frac{u(x)}{x^2}\,dx$$

$$= \frac{v(t)}{t} - \frac{1}{\varphi(f(t))}\int_t^b \varphi\left(\left|f(t) - \frac{1}{x}\int_0^x f(s)\,ds\right|\right)\frac{u(x)}{x^2}\,dx. \quad (5.37)$$

Therefore, (5.36) follows from (5.37) and Theorem 5.14. The proof is complete. □

Next we state a "dual" version of Theorem 5.14. Note that the natural dual operator of the Hardy operator $H : Hf(x) = \frac{1}{x}\int_0^x f(t)\,dt$ is $\hat{H} : \hat{H}f(x) = \int_x^\infty \frac{f(t)}{t}\,dt$, but here we use its alternative $H^* : H^*f(x) = x\int_x^\infty \frac{f(t)}{t^2}\,dt$.

Theorem 5.15 *Let $-\infty \leq a < c \leq \infty$, let φ be a positive and superquadratic function on (a,c), and E be a Banach function space on $[b,\infty)$, $b \geq 0$, with the Fatou property. Then, whenever $a < f(x) < c$,*

$$\left\|\varphi\left(x\int_x^\infty \frac{f(t)}{t^2}\,dt\right)\right\|_E$$

$$\leq \int_b^\infty \varphi(f(t))\left\|\left(1 - \frac{\varphi\left(|f(t) - x\int_x^\infty f(s)\frac{ds}{s^2}|\right)}{\varphi(f(t))}\right)x\chi_{[b,t]}(x)\right\|_E \frac{dt}{t^2}.$$

Proof Let $D = \{(x,t) : b \leq x, x \leq t < \infty\}$. Then

$$\chi_D(x,t) = \chi_{[x,\infty)}(t) = \chi_{[b,t]}(x). \quad (5.38)$$

5.4 Some new Hardy-type inequalities

By using (5.38) and the same arguments as in the proof of Theorem 5.14, we obtain that

$$\left\| \varphi \left(x \int_x^\infty \frac{f(t)}{t^2} \, dt \right) \right\|_E$$

$$\leq \left\| \int_x^\infty \left(\varphi(f(t)) - \varphi \left(\left| f(t) - x \int_x^\infty f(s) \frac{ds}{s^2} \right| \right) \right) x \frac{dt}{t^2} \right\|_E$$

$$= \left\| \int_b^\infty \left(\varphi(f(t)) - \varphi \left(\left| f(t) - x \int_x^\infty f(s) \frac{ds}{s^2} \right| \right) \right) x \chi_{[x,\infty)}(t) \frac{dt}{t^2} \right\|_E$$

$$= \left\| \int_b^\infty \left(\varphi(f(t)) - \varphi \left(\left| f(t) - x \int_x^\infty f(s) \frac{ds}{s^2} \right| \right) \right) x \chi_D(x,t) \frac{dt}{t^2} \right\|_E$$

$$\leq \int_b^\infty \left\| \left(\varphi(f(t)) - \varphi \left(\left| f(t) - x \int_x^\infty f(s) \frac{ds}{s^2} \right| \right) \right) x \chi_D(x,t) \right\|_E \frac{dt}{t^2}$$

$$= \int_b^\infty \left\| \left(\varphi(f(t)) - \varphi \left(\left| f(t) - x \int_x^\infty f(s) \frac{ds}{s^2} \right| \right) \right) x \chi_{[b,t]}(x) \right\|_E \frac{dt}{t^2}$$

$$= \int_b^\infty \varphi(f(t)) \left\| \left(1 - \frac{\varphi \left(\left| f(t) - x \int_x^\infty f(s) \frac{ds}{s^2} \right| \right)}{\varphi(f(t))} \right) x \chi_{[b,t]}(x) \right\|_E \frac{dt}{t^2}.$$

The proof is complete. □

We give the following example of application of Theorem 5.15 (c.f. [104, Proposition 2.2]):

Corollary 5.4 *Let* $0 \leq b < \infty$, $u : (b, \infty) \to \mathbb{R}$ *be a nonnegative locally integrable function on* (b, ∞), *and define the function* v *by*

$$v(t) = \frac{1}{t} \int_b^t u(x) \, dx, \quad t \in (b, \infty).$$

If the real-valued function φ *is positive and superquadratic on* (a, c), $0 \leq a < c \leq \infty$, *then the inequality*

$$\int_b^\infty u(x) \varphi \left(x \int_x^\infty f(t) \frac{dt}{t^2} \right) \frac{dx}{x} \qquad (5.39)$$

$$\leq \int_b^\infty v(t) \varphi(f(t)) \frac{dt}{t} - \int_b^\infty \int_b^t \varphi \left(\left| f(t) - x \int_x^\infty f(s) \frac{ds}{s^2} \right| \right) u(x) \, dx \frac{dt}{t^2}$$

holds for all f *with* $a < f(x) < c$, $x \geq b$.

Proof It is known that $E = L_1\left([b, \infty), \frac{u(x)}{x} dx\right)$ satisfies the Fatou property (see, e.g., [17]). Moreover,

$$\left\|\left(1 - \frac{\varphi\left(|f(t) - x\int_x^\infty f(s)\frac{ds}{s^2}|\right)}{\varphi(f(t))}\right) x\chi_{[b,t]}(x)\right\|_E$$

$$= \int_b^t u(x) dx - \frac{1}{\varphi(f(t))} \int_b^t \varphi\left(\left|f(t) - x\int_x^\infty f(s)\frac{ds}{s^2}\right|\right) u(x) dx$$

$$= tv(t) - \frac{1}{\varphi(f(t))} \int_t^b \varphi\left(\left|f(t) - x\int_x^\infty f(s)\frac{ds}{s^2}\right|\right) u(x) dx. \tag{5.40}$$

Therefore, (5.39) follows from (5.40) and Theorem 5.15, so the proof is complete. □

The next two results can be considered as a refinement of the Hardy inequality via a strongly convex function.

Theorem 5.16 *Let $0 < b \leq \infty, -\infty \leq a < d \leq \infty$, let φ be a positive and strongly convex with modulus $c > 0$ function on (a, d), and E be a Banach function space on $[0, b)$. If E has the Fatou property and $a < f(x) < d$, then*

$$\left\|\varphi\left(\frac{1}{x}\int_0^x f(t) dt\right)\right\|_E$$

$$\leq \int_0^b \varphi(f(t)) \left\|\left(1 - \frac{c\left(f(t) - \frac{1}{x}\int_0^x f(s) ds\right)^2}{\varphi(f(t))}\right) \frac{1}{x}\chi_{[t,b)}(x)\right\|_E dt,$$

provided that both sides have sense.

The proof of this theorem can be done in a similar way as the proof of Theorem 5.14. We give the proof just for the completeness.

Proof Let $D = \{(x, t) : 0 \leq x \leq b, 0 \leq t \leq x\}$. Then (5.35) holds. By using Theorem 3.1, the lattice property of E, Theorem 5.5 for $p = 1$ and (5.35), we find that

$$\left\|\varphi\left(\frac{1}{x}\int_0^x f(t) dt\right)\right\|_E \leq \left\|\int_0^x \frac{\varphi(f(t)) - c\left(f(t) - \frac{1}{x}\int_0^x f(s) ds\right)^2}{x} dt\right\|_E$$

$$= \left\|\int_0^b \frac{\varphi(f(t)) - c\left(f(t) - \frac{1}{x}\int_0^x f(s) ds\right)^2}{x} \chi_{[0,x]}(t) dt\right\|_E$$

5.4 Some new Hardy-type inequalities

$$= \left\| \int_0^b \frac{\varphi(f(t)) - c\left(f(t) - \frac{1}{x}\int_0^x f(s)\,ds\right)^2}{x} \chi_D(x,t)\,dt \right\|_E$$

$$\leq \int_0^b \left\| \frac{\varphi(f(t)) - c\left(f(t) - \frac{1}{x}\int_0^x f(s)\,ds\right)^2}{x} \chi_D(x,t) \right\|_E dt$$

$$= \int_0^b \left\| \frac{\varphi(f(t)) - c\left(f(t) - \frac{1}{x}\int_0^x f(s)\,ds\right)^2}{x} \chi_{[t,b]}(x) \right\|_E dt$$

$$= \int_0^b \varphi(f(t)) \left\| \left(1 - \frac{c\left(f(t) - \frac{1}{x}\int_0^x f(s)\,ds\right)^2}{\varphi(f(t))}\right) \frac{1}{x} \chi_{[t,b]}(x) \right\|_E dt.$$

The proof is complete. □

Here we just give one example of application of Theorem 5.14 (c.f. [104, Proposition 2.1] and [54, Theorem 2.3]):

Corollary 5.5 *Let $0 < b \leq \infty, u : (0,b) \to \mathbb{R}$ be a nonnegative weight function such that the function $x \mapsto \frac{u(x)}{x^2}$ is locally integrable on $(0,b)$, and define the weight function v by*

$$v(t) = t \int_t^b \frac{u(x)}{x^2}\,dx, \quad t \in (0,b).$$

If the function φ is positive and strongly convex with modulus $c > 0$ on $(a,d), 0 \leq a < d \leq \infty$, then the inequality

$$\int_0^b u(x)\,\varphi\left(\frac{1}{x}\int_0^x f(t)\,dt\right) \frac{dx}{x} \tag{5.41}$$

$$\leq \int_0^b v(t)\varphi(f(t))\,\frac{dt}{t} - c\int_0^b \int_t^b \left(f(t) - \frac{1}{x}\int_0^x f(s)\,ds\right)^2 \frac{u(x)}{x^2}\,dx\,dt$$

holds for all f with $a < f(x) < d, 0 < x \leq b$.

Proof It is known that $E = L_1\left((0, b), \frac{u(x)}{x}\,dx\right)$ satisfies the Fatou property. Moreover,

$$\left\|\left(1 - \frac{c\left(f(t) - \frac{1}{x}\int_0^x f(s)\,ds\right)^2}{\varphi(f(t))}\right)\frac{1}{x}\chi_{[t,b]}(x)\right\|_E$$

$$= \int_t^b \frac{u(x)}{x^2}\,dx - \frac{c}{\varphi(f(t))}\int_t^b \left(f(t) - \frac{1}{x}\int_0^x f(s)\,ds\right)^2 \frac{u(x)}{x^2}\,dx$$

$$= \frac{v(t)}{t} - \frac{c}{\varphi(f(t))}\int_t^b \left(f(t) - \frac{1}{x}\int_0^x f(s)\,ds\right)^2 \frac{u(x)}{x^2}\,dx. \qquad (5.42)$$

Therefore, (5.41) follows from (5.42) and Theorem 5.16. The proof is complete. □

Remark 5.8 Some "dual" forms of Theorem 5.16 and Corollary 5.5 also hold. The formulations and proofs are completely analogous as in superquadratic case (see Theorem 5.15 and Corollary 5.4, respectively) so we leave the details to the reader.

We finish this chapter by stating and proving a Bellman-type inequality involving the Hardy operator which is defined as

$$Hf(y) = \frac{1}{y}\int_0^y f(t)\,dt.$$

We recall that a Banach function space E on the half line $(0, \infty)$ is said to be symmetric if it has the following property: If $y \in E$ and for the nonincreasing rearrangement invariant functions, we have that $x^*(t) \leq y^*(t)$ for all $t \in (0, \infty)$, then $x \in E$ and $\|x\|_E \leq \|y\|_E$. Moreover, if E is a symmetric Banach function space, then for a dilatation operator $\sigma_\tau f(t) = f(\tau^{-1}t)$, $\tau > 0$, the following holds: $\|\sigma_\tau\|_E \leq \max\{1, \tau\}$ (see [59]).

Applying the Bellman-type inequality given in Theorem 5.9, we get the following result (see [100]):

Theorem 5.17 *Let $p \geq 1$, $\alpha \in \mathbb{R}$ and $f_0(x)$ be a function on $(0, 1)$ such that $f_0^p(x) > \frac{1}{x^{\alpha+1}}\|f^p(y)y^\alpha\|_E$, where $E = E(0, \infty)$ is a symmetric Banach function space on $Y = (0, \infty)$ for all $x \in (0, 1)$. Assume that E has the Fatou property. Then the inequality*

$$\|(Hf)^p(y)y^\alpha\|_E \leq \left(\int_0^1 f_0(x)\,dx\right)^p - \left(\int_0^1 \left[f_0^p(x) - \|f^p(xy)y^\alpha\|_E\right]^{\frac{1}{p}}\,dx\right)^p$$

$$\leq \left(\int_0^1 f_0(x)\,dx\right)^p - \left(\int_X \left[f_0^p(x) - \frac{1}{x^{\alpha+1}}\|f^p(y)y^\alpha\|_E\right]^{\frac{1}{p}}\,dx\right)^p$$

holds.

5.4 Some new Hardy-type inequalities

Proof Put in inequality (5.19) $X = (0, 1)$, $Y = (0, \infty)$ and instead of function $f(x, y)$ consider a function $g(x, y) = f(xy)y^{\alpha/p}$. Then

$$\|g^p(x, y)\|_E = \|f^p(xy)y^\alpha\|_E$$

$$= \frac{1}{x^\alpha}\|f^p(xy)(xy)^\alpha\|_E = \frac{1}{x^\alpha}\|\sigma_{x^{-1}}(f^p(y)y^\alpha)\|_E$$

$$\leq \frac{1}{x^\alpha}\|\sigma_{x^{-1}}\| \cdot \|f^p(y)y^\alpha\|_E \leq \frac{1}{x^\alpha}\max\{1, x^{-1}\}\|f^p(y)y^\alpha\|_E$$

$$= \frac{1}{x^{\alpha+1}}\|f^p(y)y^\alpha\|_E.$$

On the other hand the change of variables $t = xy$ implies that

$$\|(Hf)^p(y)y^\alpha\|_E = \left\|\left(\frac{1}{y}\int_0^y f(t)\,dt\right)^p y^\alpha\right\|_E = \left\|\left(\frac{1}{y}\int_0^1 f(xy)y^{\frac{\alpha}{p}}y\,dx\right)^p\right\|_E$$

$$= \left\|\left(\int_0^1 g(x, y)\,dx\right)^p\right\|_E.$$

Hence, by Theorem 5.9 the proof is complete. $\square$

If $E = L_1(Y)$, we get the following result (see [100]):

Corollary 5.6 *Let $p \geq 1$ and $\alpha \in \mathbb{R}$. Then the inequality*

$$\int_0^\infty (Hf)^p(y)y^\alpha\,dy$$

$$\leq \left(\int_0^1 f_0(x)\,dx\right)^p - \left(\int_0^1 \left[f_0^p(x) - \int_0^\infty f^p(xy)y^\alpha\,dy\right]^{\frac{1}{p}}dx\right)^p$$

holds for any positive function $f(x)$ whenever the function $f_0(x)$ satisfies $f_0^p(x) > \int_0^\infty f^p(xy)y^\alpha\,dy$ or, which is the same, $f_0^p(x) > \frac{1}{x^{\alpha+1}}\int_0^\infty f^p(s)s^\alpha\,ds$ for all $x \in (0, 1)$, provided that all integrals exist.

Remark 5.9 We note that the Hardy inequality (2.37) holds only under the restriction $\alpha < p - 1$, i.e., we can estimate the quantity $\int_0^{d_0}\left(\frac{1}{x}\int_0^x g(y)\,dy\right)^p x^\alpha\,dx$ only under this restriction. However, according to Corollary 5.6 this quantity can be estimated also without this restriction if we instead put the following restriction on g:
There exists a function $f_0(x)$ such that $f_0^p(x) > \frac{1}{x^{\alpha+1}}\int_0^\infty g^p(y)y^\alpha\,dy$.

Appendix A

On the close relation between inequalities, interpolation theory, and convexity. Interpolation between families of Banach spaces vis-à-vis continuous inequalities

Here we give some information about interpolation theory in the classical situation of two spaces, in finite and infinite families of Banach spaces, where some of us worked till the early 2000s, and try to show how this attracted our attention to continuous forms of inequalities.

Interpolation theory usually concerns interpolation between two Banach spaces A_0 and A_1. See, for example, the monographs [17, 18, 22, 59, 122] and the references therein. To quickly understand the basic idea and close relation between inequalities, interpolation theory, and convexity, we present the following classical Riesz-Thorin interpolation theorem where $A_0 = L_{p_0}$ and $A_1 = L_{p_1}$.

Theorem A.1 *Let* $p_0 \neq p_1, q_0 \neq q_1$, $p_i, q_i \geq 1$. *Let*

$$T : L_{p_0} \to L_{q_0} \text{ with norm } M_0,$$

i.e., the inequality

$$\|Tf\|_{L_{q_0}} \leq M_0 \|f\|_{L_{p_0}} \qquad (A.1)$$

holds and

$$T : L_{p_1} \to L_{q_1} \text{ with norm } M_1$$

i.e., the inequality

$$\|Tf\|_{L_{q_1}} \leq M_1 \|f\|_{L_{p_1}} \tag{A.2}$$

holds.

Let

$$\frac{1}{p_\theta} = \frac{1-\theta}{p_0} + \frac{\theta}{p_1}, \quad \frac{1}{q_\theta} = \frac{1-\theta}{q_0} + \frac{\theta}{q_1}, \quad \theta \in (0,1). \tag{A.3}$$

Then

$$T : L_{p_\theta} \to L_{q_\theta} \text{ with norm } M_\theta$$

and

$$M_\theta \leq M_0^{1-\theta} M_1^\theta, \tag{A.4}$$

i.e., the inequality

$$\|Tf\|_{L_{q_\theta}} \leq M_0^{1-\theta} M_1^\theta \|f\|_{L_{p_\theta}} \tag{A.5}$$

holds.

Inequality (A.4) means that $\log M$ is a convex function, (A.3) is an interpolation property between the parameters $\frac{1}{p}$ and $\frac{1}{q}$, respectively, and the inequalities (A.1) and (A.2) imply that the "interpolated" inequality (A.5) between L_p spaces holds. Already this theorem is very powerful for improving various inequalities in analysis. The original motivation to prove this theorem was to get a simple proof of the Hausdorff-Young famous inequality. The original proof of this inequality was very long and fairly complicated, but by using this new interpolation theory it can be done in one line.

We have already gotten a flavor of the relationship between inequalities, interpolation, and convexity, i.e., we see that M_θ is log convex. Moreover, the Hölder inequality, also called the Littlewood inequality in this concrete situation,

$$\|f\|_{L_{p_\theta}} \leq \|f\|_{L_{p_0}}^{1-\theta} \|f\|_{L_{p_1}}^\theta \tag{A.6}$$

is an inequality of log-convex type.

The close relation between convexity and inequalities is also well-known and used frequently by G. H. Hardy himself, e.g., in the monograph [47]. Concerning various aspects of convexity, including the connection to inequalities, we refer to the monograph [85].

A Appendix

Just how this works, we mention that the classical Hardy inequality

$$\int_0^\infty \left(\frac{1}{x}\int_0^x f(y)\mathrm{d}y\right)^p x^a \mathrm{d}x \leq \left(\frac{p}{p-1-a}\right)^p \int_0^\infty f^p(y) y^a \mathrm{d}y, \tag{A.7}$$

where $a < p-1$, $p \geq 1$, is via the substitution $f(x) = g(x^{(p-1-a)/p})x^{\frac{-1-a}{p}}$ equivalent to the inequality

$$\int_0^\infty \left(\frac{1}{x}\int_0^x g(y)\mathrm{d}y\right)^p \frac{\mathrm{d}x}{x} \leq \int_0^\infty g^p(x)\frac{\mathrm{d}x}{x},$$

and this inequality follows directly from convexity in the form of the Jensen inequality (see Remark 2.5). The original proof of (A.7) took long time to find, but by using this fairly new convexity relation the proof will again be one line. For some further consequences of this convexity approach, see [114]. For an "elementary" description of this close relation between convexity, inequalities, and interpolation, we refer to [113].

This book is devoted to continuous forms of classical inequalities. There is also a less-known theory concerning interpolation between families of (continuously many) Banach spaces instead of two Banach spaces as in the classical situation. Also, in these more general situations, there are many similarities and relations between continuous inequalities, interpolation between families of Banach spaces, and (generalized) convexity. Indeed, in our works concerning interpolation of families of Banach spaces (see [29, 31, 57, 58, 88–90]) we first understood the importance of proving continuous inequalities (mostly of Hölder type). See also [27, 28, 30]. And moreover, to understand the close connection between these subjects and (generalized) types of convexity.

There are several methods concerning interpolation, e.g., the real, complex, and Calderón methods. And these three methods have been developed to cover also the more general case which deals with interpolation between families of Banach spaces. Just to give the reader a flavor of this wonderful interpolation theory, we shall first give some basic definitions related to the complex method of interpolation and also Calderón's method concerning interpolation between two Banach lattices X_0 and X_1 on a measure space.

Consider Banach couples, i.e., pairs (A_0, A_1) such that A_0 and A_1 are Banach spaces embedded in a common topological vector space U. Very important among the various constructions of interpolation concerning to a given couple are the complex method leading to the spaces $[A_0, A_1]_\theta$ (where $0 < \theta < 1$) and the real method leading to the spaces $(A_0, A_1)_{\theta,q}$ (where $0 < \theta < 1$ and $1 \leq q \leq \infty$). Usually, the Banach couple is denoted by $\overline{A} = (A_0, A_1)$, the intersection $\Delta \overline{A} = A_0 \cap A_1$ is provided with the norm $\|a\|_{\Delta \overline{A}} = \max(\|a\|_{A_0}, \|a\|_{A_1})$, the sum $\Sigma \overline{A} = A_0 + A_1$ consists of those elements a of U, which can be represented as $a = a_0 + a_1$, $a_0 \in A_0$, $a_1 \in A_1$, and its norm is $\|a\|_{A_0+A_1} = \|a\|_{\Sigma \overline{A}} = \inf_{a=a_0+a_1}(\|a_0\|_{A_0} + \|a_1\|_{A_1})$.

We say that A and B are interpolation spaces of type θ if quasi-log-convexity inequality

$$\|T\|_{A\to B} \leq C(\|T\|_{A_0\to B_0})^{1-\theta}(\|T\|_{A_1\to B_1})^\theta$$

holds. If $C = 1$, i.e., if we have log-convexity inequality, then we say that A and B are exact interpolation spaces of type θ.

Let us briefly describe the complex method. Consider functions, defined in the strip $\Pi = \{z : 0 < \operatorname{Re} z < 1\}$ with values in Banach spaces. Let A be a complex Banach space. Let $F(A)$ denote the set of functions $f : \overline{\Pi} \to A$, bounded and continuous in $\overline{\Pi}$, holomorphic in Π.

Let $\overline{A} = (A_0, A_1)$ be a Banach couple of complex spaces. $F(A_0, A_1)$ denotes the set of functions $f : \overline{\Pi} \to A$ with the following properties:

(1) $f(z)$ is continuous and bounded by the norm of $A_0 + A_1$.
(2) $f(z)$ is holomorphic by the norm of $A_0 + A_1$ in Π.
(3) $f(it)$ is continuous and bounded by the norm of A_0, and $f(1 + it)$ is continuous and bounded by the norm of A_1.

Put

$$\|f\|_{F(A_0, A_1)} = \max\{\sup\|f(it)\|_{A_0}, \sup\|f(1+it)\|_{A_1}\}.$$

Let $0 \leq \theta \leq 1$. The interpolation space of the complex method is defined as follows:

$$[A_0, A_1]_\theta = \{a \in A_0 + A_1; a = f(\theta) \text{ for some } f \in F(A_0, A_1)\}$$

with

$$\|a\|_{[A_0, A_1]_\theta} = \inf_{f(\theta)=a, f\in F(A_0,A_1)} \|f\|_{F(A_0,A_1)}.$$

Note that this method gives exact interpolation spaces of type θ. In particular,

$$\|f\|_{[A_0,A_1]_\theta} \leq \|f\|_{A_0}^{1-\theta} \|f\|_{A_1}^\theta, \tag{A.8}$$

and when $A_0 = L_{p_0}, A_1 = L_{p_1}$, we get $[A_0, A_1]_\theta = L_{p_\theta}$ and, hence, (A.8) means that the complex method is an exact interpolation method and the generalized Hölder inequality (A.6) holds.

Next, we shortly describe the Calderón method of interpolation for the case of two Banach lattices (see [26]). This is a simply looking definition of interpolation spaces. Here we consider the case of a couple of Banach lattices X_0 and X_1 on the space M with σ-finite measure μ. A more general is the Calderón-Lozanowsky method.

Fix $\theta \in (0, 1)$ and denote by X the space of all complex-valued functions $x(s)$ on Ω such that the quasi-log-convexity inequality

$$|x(s)| \leq \lambda |x_0(s)|^{1-\theta} |x_1(s)|^{\theta} \qquad (A.9)$$

holds for some $\lambda > 0$, $x_0 \in X_0$, $x_1 \in X_1$ with $\|x_0\|_{X_0} \leq 1$, $\|x_1\|_{X_1} \leq 1$. This is a linear space that becomes a Banach space (even a Banach lattice) with the norm $\|x\|_X = \inf \lambda$ where the inf is taken over all λ for which there exist x_0 and x_1 satisfying inequality (A.9). It is usually denoted by $X_0^{1-\theta} X_1^{\theta}$.

These basic definitions lead to the fact that an interpolation theorem, similar to the Riesz-Thorin Theorem A.1, holds (A_0, A_1, and A_θ correspond to L_{p_0}, L_{p_1}, and L_{p_θ}, respectively).

The appearance of a term in form of $a^{1-\theta} b^{\theta}$ in all three inequalities (A.4), (A.6) and (A.9) makes us to think about connection between them.

A suggestion of extending the complex interpolation method to n Banach spaces was made by J. L. Lions in 1960 (see [74]) and studied in detail by A. Favini in 1972 (see [41]). In his work, A. Favini studied the complex interpolation method for three Banach spaces, and the extension from three to n Banach spaces does not present any new essential difficulties. Shortly, instead of the strip $\overline{\Pi}$, the set $\Omega = \{(z, w) : \text{Re } z \geq 0, \text{Re } w \geq 0, 0 \leq \text{Re } z + \text{Re } w \leq 1\}$ and holomorphic functions $f(z, w)$ are used.

Theory of interpolation of finite families $A_0, A_1, \ldots, A_n$ of Banach spaces began to be considered after the year 1970, for the complex method by A. Favini 1972 and J. L. Lions (see [41, 74]) and L. Nikolova 1978 (see [87]), for the real method—G. Sparr 1974 (see [121]), D. L. Fernandez 1979 (see [42]) and later by F. Cobos and J. Peetre (see [33]).

What happens when we "run to infinity" (quotation of prof. S. Krein's expression), namely when we consider interpolation between an infinite family of Banach spaces? The theory of interpolation in infinite families began its development in about 1980—the complex interpolation method and later the real interpolation method, and also the Calderón method in the case of Banach lattices.

The continuous form of inequalities (again infinite case)—the Hölder, Minkowski, and so on, in some way corresponds to the infinite case, and this answers the above question of how interpolation of families of Banach spaces inspires interest in the continuous form of classical inequalities.

We will shortly describe the complex interpolation method of an infinite family of Banach spaces and the Calderón method in the case of an infinite family of Banach lattices and also give information of some kind of real interpolation in such infinite families.

Parts of the theory concerning interpolation between two Banach spaces can be generalized to cover also, the cases where one interpolates between finitely many Banach spaces and even between general families of (infinitely many) Banach spaces.

When speaking about interpolation of (infinite) families of Banach spaces, at first, we have to mention the complex methods (see [34, 35, 57, 58]), which appeared before the real methods. In terms of [34], let D be a suitable simple connected domain in the complex plane with boundary Γ and $A(\gamma) \in \Gamma$ be an interpolation family on Γ in the sense of [34]. Let for simplicity $\Gamma = \{|z| = 1\}$, $D = \{|z| < 1\}$. When we speak about interpolation in the families of Banach spaces (complex or real), we are in the situation when the actual family of Banach spaces is indexed by the points of the unit circle $\Gamma = \{|z| = 1\}$ in the complex plane, while the interpolation spaces are labeled by the points of the unit disk $D = \{|z| < 1\}$. The family $\bar{A} = \{A(\gamma), \gamma \in \Gamma\}$ of Banach spaces is called an interpolation family if each $A(\gamma)$ is continuously embedded in a large Banach space U and if the function $\gamma \to \|x\|_{A(\gamma)}$ is measurable for each $x \in \cap_{\gamma \in \Gamma} A(\gamma)$ and if $\|x\|_U \leq K(\gamma)\|x\|_{A(\gamma)}$ for all x from β (the log-intersection space of the family), where

$$\beta = \left\{ x \in \cap_{\gamma \in \Gamma} A(\gamma), \int_0^{2\pi} \log^+ \|x\|_{A(\gamma)}(\gamma) dt < \infty \right\}$$

and $\log^+ K(t) \in L_1(\Gamma)$.

Let $N^+(A(\gamma))$ denote the space of all β-valued analytic functions of the form $g(z) = \Sigma \psi_j(z) x_j$, where $\psi_j \in N^+$ (N^+ denotes Nevanlinna class on the unit disk D), for which $\|g\|_\infty = \sup \|g(\gamma)\|_{A(\gamma)} < \infty$. The space $A(z), z \in D$ consists of all elements x of the form $f(z)$ for $f \in F$, where F denotes the completion of $N^+(X(t))$ with respect to the above norm $\|.\|_\infty$. In $A(z)$ a norm $\|.\|_{A(z)}$ is defined by the formula:

$$\|x\|_{A(z)} = \inf\{\|f\|_\infty, f \in F, f(z) = x\}.$$

A variant of this construction was suggested independently in [57]. The interpolation theorem reads (see [34]):

Theorem A.2 *Let T be a linear operator which maps U continuously into V, where U and V contain spaces of the families $\bar{A} = \{A(\gamma), \gamma \in \Gamma\}$ and $\bar{B} = \{B(\gamma), \gamma \in \Gamma\}$, respectively. Suppose further that T maps $\{A\}$ (the log-intersection space of the family $\{A(\gamma)\}$) into $\cap_{\gamma \in \Gamma} B(\gamma)$ with $\|Ta\|_{B_\gamma} \leq M(t)\|a\|_{A_\gamma}$ for all $a \in \{A\}, t \in T$, where $\log M(\gamma)$ is absolutely integrable on Γ. Then, T maps $A(z)$ into $B(z)$ with norm not exceeding*

$$M(z) = \exp\left(\int_\Gamma \log M(\gamma) P_z(\gamma) d\gamma\right),$$

where P_z is the Poisson kernel.

The estimate of $M(z)$ can be regarded as an infinite variant of the inequality (log-convexity inequality) in the notion of the exact interpolation method of type θ (here z) in the case when the families consist of just two spaces (the case of Banach couples).

A Appendix

Proposition 2.4 from [34] says that for each $f \in F(\gamma)$ and each $z_0 \in D$

$$\|f(z_0)\|_{A(z_0)} \leq \exp\left(\int_\Gamma \log \|f(\gamma)\|_{A(\gamma)} \, dP_{z_0}(\gamma)\right) \quad (A.10)$$

where $P_{z_0}(\gamma)$ being the Poisson kernel.

Let us note the following result from [34] about interpolation of a family of Lebesgue spaces on a measure space (X, μ):

Let $p(\gamma)$ be a measurable function on Γ with $1 \leq p(\gamma) < \infty$. If $A(\gamma) = L_{p(\gamma)}$, then $A(z_0) = L_{p(z_0)}$, where $1/p(z_0) = \int_\Gamma \frac{1}{p(\gamma)} \, dP_{z_0}(\gamma)$.

Remember that a Banach lattice X has the dominated convergence property if, given $f \in X$ and $\{f_n\}_{n=1}^\infty$, such that $f_n \leq f, n = 1, 2, \ldots$ and $f_n \to 0$ as $n \to \infty$, then $\|f_n\|_X \to 0$ as $n \to \infty$. In the following, $X(\gamma)$ is a Banach lattice (Banach function space) on the measure space (Y, Σ, ν). A family of Banach lattices $X(\gamma), \gamma \in \Gamma$, is called an interpolation family if it is an interpolation family of Banach spaces for which the containing space is also a Banach lattice and $\|F(y, \gamma)\|_{X(\gamma)}$ is a measurable function of γ for all measurable $F : Y \times T \to R$ is such that $F(., \gamma) \in X(\gamma)$ a.e. We are going to define the Calderón interpolation space $[X(\gamma)]^z, z \in D$, for an infinite family suggested by E. Hernandez in [48]. As for notation he systematically used the letter γ instead of $e^{i\gamma}$ to denote an element of $\Gamma = \{z \in C, |z| = 1\}$.

Let now the family $\bar{X} = \{X(\gamma), \gamma \in \Gamma\}$ be a family of Banach lattices on a fixed measure space (M, ds). Define the space $[X(\gamma)]^z, z \in D$, which consists of all measurable functions f on (M, ds), for which there exist $\lambda > 0$ and a measurable function $F : M \times \Gamma \to R$ with $\|F(., \gamma)\|_{X(\gamma)} \leq 1$ a.e. such that

$$|f(s)| \leq \lambda \exp\left(\int_\Gamma \log |F(s, \gamma)| P_z(\gamma) \, d\gamma\right).$$

The norm $\|f\|_{[X(\gamma)]^z}$ is defined to be the infimum of the values of λ for which such an inequality holds.

Next we formulate a variant of Theorem 6.1 from [48] just for simplicity for the case $B(\gamma) = \mathbb{R}$.

Theorem A.3 *Suppose that $\{X(\gamma), \gamma \in \Gamma\}$, is an interpolation family of Banach lattices and $X(z)(= [X(\gamma)]_z)$ is the complex interpolation space constructed for this family. Then $X(z) \subset [X(\gamma)]^z$. If, in addition, we assume that $[X(\gamma)]^z$ has the dominated convergence property, then the spaces $[X(\gamma)]^z$ and $X(z)$ coincide and their norms are equal.*

It follows from this that if $[X(\gamma)]^z$ has the dominated convergence property, then for $z \in D, [X(\gamma)]^z = X(z)$.

The following Lemma was used in [48] in the proof of the theorem of interpolation of Lorentz spaces. Remember that when defining Lorentz space, we need the rearrangement invariant f^* of a function f and also f^{**}. Here we mention the definition $f^{**}(t) = \frac{1}{t} \sup \int_e f(s) \, d\mu$ where the supremum is taken over all measurable sets in M with $\mu(e) \leq t$.

Lemma A.1 *Let $F : M \times \Gamma \to \mathbb{R}$ be a measurable function with*

$$\int_T \log \|F(., \gamma)\|_{L_1} P_z(\gamma) \, d\gamma < \infty$$

for $z \in D$. Then

$$\left\{ \exp \left(\int_T \log F(., \gamma) P_z(\gamma) \, d\gamma \right) \right\}^{**}(t) \leq \exp \left(\int_T \log F(t, \gamma)^{**} P_z(\gamma) \, d\gamma \right).$$

Next, we will briefly describe a variant of the real method of interpolation in families of Banach spaces. An extension of the real interpolation method to a finite number of Banach spaces was made by G. Sparr (see [121]), A. Yoshikawa (see [127]), D. L. Fernandez (see [42]) and later F. Cobos and J. Peetre (see [33]). The construction of G. Sparr was extended by M. Cwikel and S. Janson in [35]. Concerning real interpolation in a finite family, let us mention I. Asekritova and N. Krugljak (see [8]) and I. Asekritova et al. (see [9]). In the classical theory of real interpolation, one basic notation is the (Peetre) K-functional, which was introduced by J. Peetre.

For simplicity, we shall work with bounded families, i.e., we assume that $\|a\|_U \leq \|a\|_{A(\gamma)}$. For each $a \in U$ we define the discrete K-functional $K(\alpha, a)$ as follows:

$$K(\alpha, a) = \inf \left\{ \sum_j \alpha(\gamma_j) \|a_{\gamma_j}\|_{A_{\gamma_j}} \right\},$$

where the infimum is taken over all representations of the element $a = \sum_j a_{\gamma_j}$ with convergence in U and $a_{\gamma_j} \in A(\gamma_j)$.

Let $S \subset L = \{\alpha : \Gamma \to \mathbb{R}^+, \log \alpha \in L_1(\Gamma)\}$. We will suppose that it is a multiplicative subgroup, bounded from above and below by positive constant functions.

In the sequel we use notation $\alpha(z) = \exp \left(\int_\Gamma \log \alpha(\gamma) P_z(\gamma) d\gamma \right)$, where, as usual, P_z denotes the Poisson kernel.

For $1 \leq p \leq \infty$, we define the following interpolation space $(A)^S_{z_0, p; K}$ in the following way:

$$(A)^S_{z_0, p; K} = \left\{ a \in U \, ; \, \left(\sum_{\alpha \in S} \left(\frac{K(\alpha, a)}{\alpha(z_0)} \right)^p \right)^{1/p} < +\infty \right\},$$

with the norm $\|a\|_{(A)^S_{z_0, p; K}} = \left(\sum_{\alpha \in S} \left(\frac{K(\alpha, a)}{\alpha(z_0)} \right)^p \right)^{1/p}$.

A Appendix

We say that a Banach space A belongs to the class $K_{z_0}^S(\bar{A})$ iff $A \subset \Sigma A_\gamma (= \{a \in U : K(1, a) < \infty\})$ and for any $a \in S$ bounded from above and below by positive constants and the following inequality holds

$$K(\alpha, a) \leq C \exp\left(\int_\Gamma \log \alpha(\gamma) \, P_z(\gamma) d\gamma\right).$$

Note that if $\sum_{\alpha \in S} \frac{\text{ess inf} \alpha(\gamma)}{\alpha(z_0)} < \infty$, then the spaces $(A)_{z_0, p; K}^S$ belong to the class $K_{z_0}^S(\bar{A})$. Moreover, the complex interpolation space $A(z_0)$ also belongs to that class.

Moreover, the related J-functional and the corresponding interpolation space were defined and studied in [8] and [9].

In [88] L. Nikolova considered a discrete K-functional for $q = 1$, while the continuous form of a K-functional was defined and investigated by M. J. Carro in 1994 (see [27]). When speaking about the real method, we have to mention Sparr's method, Fernandez's method, and Cobos-Peetre's methods related to the case of finite family, and for the infinite family the "continuous" method, suggested by M. J. Carro and the "discrete" method suggested by L. E. Persson and L. Nikolova. Note that by the suggestion of J. Peetre to join our efforts after a visit of the group in Luleå in 1997, the paper [31] by J. Peetre, M. J. Carro, L. E. Persson, and L. Nikolova appeared.

There is also a variant of a real interpolation space between a family of Banach spaces with the K-functional $K_q(\alpha, a)$ defined by

$$K_q(\alpha, a) = \inf\left\{\left(\int_\Gamma (\alpha(\gamma) \|a(\gamma)\|_\gamma)^q \, d\gamma\right)^{1/q}\right\},$$

where the infimum extends over all representations $a = \int_\Gamma a(\gamma) d\gamma$ (convergence in U) with $a(\cdot) \in \bar{G}$, where $\bar{G} = \{b = \sum_{\text{finite}} b_j \chi E_j : b_j \in A, E_j \text{ pairwise disjoint measurable sets in } \Gamma\}$ (see [28, 31]).

For $1 \leq q \leq \infty$ the space $[A]_{z_0, p, q; K}^S$ and the class $K_{z_0, q}^S(\bar{A})$ are defined in the same way like $(A)_{z_0, p; K}^S$ by just replacing the $K(\alpha, a)$ by $K_q(\alpha, a)$. Again the spaces $(A)_{z_0, p, q; K}^S$ and the complex interpolation space $A(z_0)$ belong to the class $K_{z_0, q}^S(\bar{A})$.

It is known (see [29, 30]) that for the K-method the weakly compactness and the compactness properties are preserved by interpolation for every $q > 1$ but not in the case $q = 1$. Results about interpolation of properties of operators being compact, weakly compact, limited, order summing, with Bergh-condition, C-subadditive, etc., in families of Banach spaces are given in [29, 30, 33, 88, 89, 91].

Convexity and smoothness in real interpolation spaces was investigated in [70] and [93].

When speaking about interpolation theory we also want to mention some papers about extrapolation theory. Following earlier work by Marcinkiewicz, Titchmarsh, Yano and others, a general theory of extrapolation was developed by B. Jawerth and M. Milman in [53]. For a (again infinite) family $(A_i)_{i \in \mathbb{Z}}$ of Banach spaces they defined Σ_q and Δ_q extrapolation methods for $1 < q < \infty$.

D. Edmunds and H. Triebel defined the logarithmic space $A_\theta (\log A)_{b_1, q}$ in [40]. Roughly speaking (for the case $b > 0$), for a couple of Banach spaces (A_0, A_1), A_0 being densely and continuously embedded in A_1, they constructed the complex interpolation spaces $[A_0, A_1]_\theta$ (where $0 < \theta < 1$) and then extrapolated with the Σ_q extrapolation method over $A(i) = 2^{ib}[A_0, A_1]_{\eta(i)}$ for $i \geq J$, where $J \in \mathbb{N}$ such that $\theta - 2^{-J} > 0$ and $\eta(i) = \theta - 2^{-i}$ for $i \geq J$. The usual Zygmund space $L_p(\mathrm{Log}L)_b(\Omega)$ is a particular case of such construction. In [94], [95], and [96] different properties of Edmunds-Triebel spaces are investigated.

As pronounced before, parts of the theory of interpolation between families of Banach spaces are deeply equipped with continuous inequalities (mostly of Hölder type).

In such terms we are now ready to formulate the following general Popoviciu-type inequality (see [98]).

Theorem A.4 *Let $B(\gamma), \gamma \in \Gamma$, be an interpolation family on Γ, let $f \in F(\gamma)$, $z_0 \in D$, and $B(z_0)$ be the complex interpolation space.*

If $f_0 > \|f\|_{F(\gamma)} = \operatorname{ess\,sup}_\gamma \|f(\gamma)\|_{B(\gamma)} > 0$, then

$$f_0 - \|f(z_0)\|_{B(z_0)} \geq \exp\left(\int_\Gamma \log[f_0 - \|f(\gamma)\|_{B(\gamma)}] \, dP_{z_0}(\gamma) \right).$$

Proof By using inequality (A.10) and Lemma 5.1, we find that

$$\|f(z_0)\|_{B(z_0)} + \exp\left(\int_\Gamma \log[f_0 - \|f(\gamma)\|_{B(\gamma)}] \, dP_{z_0}(\gamma) \right)$$

$$\leq \exp\left(\int_\Gamma \log \|f(\gamma)\|_{B(\gamma)} \, dP_{z_0}(\gamma) \right)$$

$$+ \exp\left(\int_\Gamma \log[f_0 - \|f(\gamma)\|_{B(\gamma)}] \, dP_{z_0}(\gamma) \right)$$

$$\leq \exp\left(\int_\Gamma \log f_0 \, dP_{z_0}(\gamma) \right) = f_0.$$

The proof is complete. □

References

1. Abramovich, S., Persson, L.-E.: Some new estimates for 'Jensen gap'. J. Inequal. Appl. **2016**, 39 (2016). https://doi.org/10.1186/s13660-016-0985-4
2. Abramovich, S., Pečarić, J., Varošanec, S.: New generalization of Gauss-Pólya's inequality. Math. Inequal. Appl. **1**(3), 331–342 (1998)
3. Abramovich, S., Jameson, G., Sinnamon, G.: Refining of Jensen's inequality. Bull. Math. Soc. Sci. Math. Roumanie (N.S.) **47(95)**(1–2), 3–14 (2004)
4. Abramovich, S., Jameson, G., Sinnamon, G.: Inequalities for averages of convex and superquadratic functions. JIPAM J. Inequal. Pure Appl. Math. **5**(4), Paper No. 91 (2004), 14 p. [electronic only] http://eudml.org/doc/124276
5. Abramovich, S., Krulić, K., Pečarić, J., Persson, L.-E.: Some new refined Hardy type inequalities with general kernels and measures. Aequationes Math. **79**(1–2), 157–172 (2010)
6. Agarwal, R.P.: Difference Equations and Inequalities. Marcel Dekker, Inc., New York (1992)
7. Ågotnes, J.J., Nikolova, L., Persson, L.-E., Varošanec, S.: Continuous inequalities: Introduction, examples and related topics. In: Duduchava, R., Shargorodsky, E., Tephnadze, G. (eds.) Tbilisi Analysis and PDE Seminar, Trends in Mathematics, Research Perspectives Ghent Analysis and PDE Center, vol. 7, pp. 1–10. Birkhäuser/Springer, Cham (2024)
8. Asekritova, I., Krugljak, N.: On equivalence of K- and J-methods for $n + 1$-tuples of Banach spaces. Studia Math. **122**(2), 99–116 (1997)
9. Asekritova, I., Krugljak, N., Maligranda, L., Nikolova, L., Persson, L.-E.: Lions-Peetre reiteration formulas for triples and their applications. Studia Math. **145**(3), 219–254 (2001)
10. Azócar, A., Nikodem, K., Roa, G.: Fejér-type inequalities for strongly convex functions. Ann. Math. Sil. **26**, 43–54 (2012)
11. Bainov, D., Simeonov, P.: Integral Inequalities and Applications. Kluwer Academic Publishers Group, Dordrecht (1992)
12. Barza, S., Nikolova, L., Persson, L.-E., Yimer, M.: Some Hardy-type inequalities in Banach function spaces. Math. Inequal. Appl. **24**(4), 1001–1016 (2021)
13. Beckenbach, E.F.: A class of mean-value functions. Amer. Math. Monthly **57**, 1–6 (1950)
14. Beckenbach, E.F., Bellman, R.: Inequalities. Springer-Verlag, Berlin (1961)
15. Beesack, P.R.: Inequalities for absolute moments of a distribution: from Laplace to von Mises. J. Math. Anal. Appl. **98**, 435–457 (1984)
16. Bellman, R.: On an inequality concerning an indefinite form. Amer. Math. Monthly **63**(2), 108–109 (1956)
17. Bennett, C., Sharpley, R.: Interpolation of Operators. Pure and Applied Mathematics 129. Academic Press, Inc., Boston (1988)

18. Bergh, J., Löfström, J.: Interpolation Spaces. An introduction. Grundlehren der Mathematischen Wissenschaften. No. 223, Springer-Verlag, Berlin (1976)
19. Bibi, R., Bohner, M., Pečarić, J., Varošanec, S.: Minkowski and Beckenbach-Dresher inequalities and functionals on time scales. J. Math. Inequal. **7**(3), 299–312 (2013)
20. Borwein, J.M., Borwein, P.B.: Pi and the AGM. Canadian Mathematical Society Series of Monographs and Advanced Texts. John Wiley & Sons, Inc., New York (1987)
21. Borwein, P.B., Erdélyi, T.: Polynomials and Polynomial Inequalities. Graduate Texts in Mathematics. Springer-Verlag, New York (1995)
22. Brudnyĭ, Yu.A., Krugljak, N.Ya.: Interpolation Functors and Interpolation Spaces, vol. 1. With a preface by Jaak Peetre. North-Holland Mathematical Library 47. North-Holland Publishing Co., Amsterdam (1991)
23. Bullen, P.S.: A Dictionary of Inequalities. Pitman Monographs Surveys in Pure and Applied Mathematics, 97. Longman, Harlow (1998)
24. Bullen, P.S.: Handbook of Means and Their Inequalities. Kluwer Academic Publishers Group, Dordrecht (2003)
25. Bullen, P.S., Mitrinović, D.S., Vasić, P.M.: Means and Their Inequalities. D. Reidel Publishing Co., Dordrecht (1988)
26. Calderón, A.P.: Intermediate spaces and interpolation, the complex method. Studia Math. **24**, 113–190 (1964)
27. Carro, M.J.: Real interpolation for families of Banach spaces. Studia Math. **109**(1), 1–21 (1994)
28. Carro, M.J.: Real interpolation for families of Banach spaces. II. Collect. Math. **45**(1), 53–83 (1994)
29. Carro, M.J., Nikolova, L.I.: Interpolation of limited and weakly compact operators on families of Banach spaces: a comparison. Acta Appl. Math. **49**(2), 151–177 (1997)
30. Carro, M.J., Peetre, J.: Some compactness results in real interpolation for families of Banach spaces. J. London Math. Soc. **58**(2), 451–466 (1998)
31. Carro, M.J., Nikolova, L.I., Peetre, J., Persson, L.-E.: Some real interpolation methods for families of Banach spaces: a comparison. J. Approx. Theory **89**(1), 26–57 (1997)
32. Cheung, W.S., Matković, A., Pečarić, J.: A variant of Jessen's inequality and generalized means. JIPAM J. Inequal. Pure Appl. Math. **7**(1), Paper No. 10 (2006), 8 p. [electronic only] http://eudml.org/doc/116837
33. Cobos, F., Peetre, J.: Interpolation of compact operators: the multidimensional case. Proc. London Math. Soc. **63**(2), 371–400 (1991)
34. Coifman, R.R., Cwikel, M., Rochberg, R., Sagher, Y., Weiss, G.: A theory of complex interpolation for families of Banach spaces. Adv. Math. **43**(3), 203–229 (1982)
35. Cwikel, M., Janson, S.: Real and complex interpolation methods for finite and infinite families of Banach spaces. Adv. Math. **66**(3), 234–290 (1987)
36. Dragomir, S.S., Khan, M.A., Abathun, A.: Refinement of the Jensen integral inequality. Open Math. **14**, 221–228 (2016)
37. Dresher, M.: Moment spaces and inequalities. Duke Math. J. **20**, 261–271 (1953)
38. Dunford, N., Schwartz, J.: Linear Operators. I. General Theory. Interscience Publishers, Inc., New York (1958)
39. Duveat, G., Lions, J.-L.: Inequalities in Mechanics and Physics. Springer-Verlag, Berlin (1976)
40. Edmunds, D., Triebel, H.: Logarithmic spaces and related trace problems. Funct. Approx. Comment. Math. **26**, 189–204 (1998)
41. Favini, A.: Su una estensione del metodo d'interpolazione complesso (Italian). Rend. Sem. Mat. Univ. Padova **47**, 243–298 (1972)
42. Fernandez, D.L.: Interpolation of 2^n Banach spaces. Studia Math. **65**, 175–201 (1979)

43. Garling, D.J.H.: Inequalities: A Journey into Linear Analysis. Cambridge University Press, Cambridge (2007)
44. Gauss, C.F.: Theoria combinationis observationum, German translation in Abhandlungen zur Methode der kleinsten Quadrate. Neudruck, Würzburg, pp. 9 and 12 (1964)
45. Guljaš, B., Pearce, C.E.M., Pečarić, J.: Some generalizations of the Beckenbach-Dresher inequality. Houston J. Math. **22**, 629–638 (1996)
46. Hardy, G.H.: Notes on some points in integral calculus, LX. An inequality between integrals. Messenger Math. **54**, 150–156 (1925)
47. Hardy, G.H., Littlewood, J.E., Pólya, G.: Inequalities. Cambridge University Press, Cambridge (1934)
48. Hernandez, E.: Intermediate spaces and the complex method of interpolation for families of Banach spaces. Ann. Scuola Norm. Sup. Pisa Cl. Sci. **13**(2), 245–266 (1986)
49. Horváth, L., Khan, K.A., Pečarić, J.: Cyclic refinements of the discrete and integral form of Jensen's inequality with applications. Analysis (Berlin) **36**, 253–262 (2016)
50. Işcan, I.: New refinements for integral and sum forms of Hölder inequality. J. Inequal. Appl. **2019**, 304 (2019). https://doi.org/10.1186/s13660-019-2258-5
51. Işcan, I.: A new improvement of Hölder inequality via isotonic linear functionals. AIMS Math. **5**(3), 1720–1728 (2020). https://doi.org/10.3934/math.2020116
52. Ivanković, B., Pečarić, J., Varošanec, S.: Properties of the Minkowski type functionals. Mediterr. J. Math. **8**(4), 543–551 (2011)
53. Jawerth, B., Milman, M.: Extrapolation theory with applications. Mem. Amer. Math. Soc. **89**(440), iv+82pp (1991)
54. Khan, M.A., Pečarić, G., Pečarić, J.: New refinement of the Jensen inequality associated to certain functions with applications. J. Inequal. Appl. **2020**, 76 (2020). https://doi.org/10.1186/s13660-020-02343-7
55. Kokilashvili, V., Krbec, M.: Weighted Inequalities in Lorentz and Orlicz Spaces. World Scientific Publishing Co., Inc., River Edge (1991)
56. Kokilashvili, V., Meskhi, A., Persson, L.-E.: Weighted Norm Inequalities for Integral Transforms with Product Kernels. Nova Science Publishers, Inc., New York (2010)
57. Krein, S.G., Nikolova, L.I.: Holomorphic functions in a family of Banach spaces, interpolation (Russian). Dokl. Akad. Nauk SSSR **250**(3), 547–550 (1980). Translation in Math. Dokl. **21**, 131–134 (1980)
58. Krein, S.G., Nikolova, L.I.: A complex interpolation method for a family of Banach spaces (Russian). Ukrain. Mat. Zh. **34**(1), 31–42 (1982). Translation in Ukrainian Math. J. **34**(1), 26–36 (1982)
59. Krein, S.G., Petunin, Ju.I., Semenov, E.M.: Interpolation of Linear Operators (Russian). Nauka, Moscow (1978)
60. Krnić, M., Lovričević, N., Pečarić, J., Perić, J.: Superadditivity and Monotonicity of the Jensen Functionals. New methods for improving the Jensen-type inequalities in real and in operator cases. Monographs in Inequalities, 11. ELEMENT, Zagreb (2015)
61. Krulić, K.: Generalizations and refinements of Hardy's inequality. Dissertation, Department of Mathematics, University of Zagreb (2010)
62. Krulić, K., Pečarić, J., Persson, L.-E.: Some new Hardy type inequalities with general kernels. Math. Inequal. Appl. **12**(3), 473–485 (2009)
63. Krulić, K., Pečarić, J., Pokaz, D.: Boas-type inequalities via superquadratic functions. J. Math. Inequal. **5**(2), 275–286 (2011)
64. Krulić Himmelreich, K., Pečarić, J., Pokaz, D.: Inequalities of Hardy and Jensen. New Hardy type inequalities with general kernels. Monographs in Inequalities, 6. ELEMENT, Zagreb (2013)

65. Kufner, A., Persson, L.-E.: Weighted Inequalities of Hardy Type. World Scientific Publishing Co., Inc., River Edge (2003)
66. Kufner, A., Maligranda, L., Persson, L.-E.: The prehistory of the Hardy inequality. Amer. Math. Monthly, **113**(8), 715–732 (2006)
67. Kufner, A., Maligranda, L., Persson, L.-E.: The Hardy Inequalities. About its history and some related results. Vydavatelsky Servis, Plzeň (2007)
68. Kufner, A., Persson, L.-E., Samko, N.: Hardy type inequalities with kernels: the current status and some new results. Math. Nach. **290**(1), 57–65 (2017)
69. Kufner, A., Persson, L.-E., Samko, N.: Weighted Inequalities of Hardy Type, 2nd edn. World Scientific Publishing Co., Pte. Ltd., Hackensack (2017)
70. Kutzarova, D., Nikolova, L.I., Zachariades, T.: Real interpolation for family of Banach spaces and convexity. Math. Nach. **171**(1), 259–268 (1995)
71. Kwon, E.G.: Extension of Hölder's inequality I. Bull. Aust. Math. Soc. **51**, 369–375 (1995)
72. Larsson, L., Maligranda, L., Pečarić, J., Persson, L.-E.: Multiplicative Inequalities of Carlson Type and Interpolation. World Scientific Publishing Co., Pte. Ltd., Hackensack (2006)
73. Lieb, E.H., Loss, M.: Analysis. Graduate Studies in Mathematics, 14. American Mathematical Society, Providence (2001)
74. Lions, J.-L.: Une construction d'espaces d'interpolation (French). C. R. Acad. Sci. Paris **251**, 1853–1855 (1960)
75. Losonczi, L., Páles, Zs.: Inequalities for indefinite forms. J. Math. Anal. Appl. **205**(1), 148–156 (1997)
76. Luxemburg, W.A.J.: On the measurability of a function which occurs in a paper by A. C. Zaanen. Netherl. Acad. Wetensch. Proc. Ser. A61 = Indag. Math. **20**, 259–265 (1958)
77. Marshall, A.W., Olkin, I.: Inequalities: Theory of Majorization and Its Applications. Academic Press, Inc., New York (1979)
78. Mercer, A.McD.: A variant of Jensen's inequality. JIPAM J. Inequal. Pure Appl. Math. **4**(4), Paper No. 73 (2003), 2 p. [electronic only] http://eudml.org/doc/123826
79. Merentes, N., Nikodem, K.: Remarks on strongly convex functions. Aequationes Math. **80**(1–2), 193–199 (2010)
80. Mitrinović, D.S.: Analytic Inequalities. Springer-Verlag, New York (1970)
81. Mitrinović, D.S., Pečarić, J., Persson, L.-E.: On a general inequality with applications. Z. Anal. Anwendungen **2**(2), 285–290 (1992)
82. Mitrinović, D.S., Pečarić, J., Fink, A.M.: Classical and New Inequalities in Analysis. Kluwer Academic Publishers Groups, Dordrecht (1993)
83. Mond, B., Pečarić, J., Šunde, J., Varošanec, S.: Inequalities of Pólya type for positive linear operators. Houston J. Math. **22**(4), 851–858 (1996)
84. Niculescu, C.P., Persson, L.-E.: Convex Functions - Basic Theory and Applications. Universitaria Press, Craiova (2003)
85. Niculescu, C.P., Persson, L.-E.: Convex Functions and Their Applications. A contemporary approach, 3rd edn. CMS Books of Mathematics. Springer, Cham (2025). First Edition 2006, Second Edition 2018
86. Nikodem, K., Páles, Zs.: Characterizations of inner product spaces by strongly convex functions. Banach J. Math. Anal. **5**(1), 83–87 (2011)
87. Nikolova, L.I.: The complex interpolation method for several spaces (Russian). C. R. Acad. Bulgar. Sci. **31**(11), 1523–1526 (1978)
88. Nikolova, L.I.: The classes K_θ and J_θ in the case of interpolation in a family of Banach spaces (Russian). C. R. Acad. Bulgare Sci. **41**(3), 9–12 (1988)
89. Nikolova, L.: Interpolation of some properties of operators acting in families of Banach spaces. Annuaire Univ. Sofia Fac. Math. Inform. **84**(1–2), 3–16 (1990)

90. Nikolova, L.I., Persson, L.-E.: Some properties of X^p-spaces. In: Function Spaces (Poznan, 1989), Teubner-Texte Mathematics, vol. 120, pp. 174–185. Teubner, Stuttgart (1991)
91. Nikolova, L.I., Persson, L.-E.: Interpolation of nonlinear operators between families of Banach spaces. Math. Scand. **72**(19), 47–60 (1993)
92. Nikolova, L., Varošanec, S.: Continuous forms of Gauss-Pólya type inequalities involving derivatives. Math. Inequal. Appl. **22**(4), 1385–1395 (2019)
93. Nikolova, L.I., Zachariades, T.: Convexity and smoothness of real interpolation spaces. Rocky Mount. J. Math. **29**(3), 1085–1101 (1999)
94. Nikolova, L.I., Zachariades, T.: The uniform convexity of the Edmunds-Triebel logarithmic spaces. J. Math. Anal. Appl **283**(2), 549–556 (2003)
95. Nikolova, L., Zachariades, T.: On Edmunds-Triebel spaces. Banach J. Math. Anal. **4**(1), 146–158 (2010)
96. Nikolova, L.I., Persson, L.-E., Zachariades, T.: On Clarkson's inequality, type and cotype for the Edmunds-Triebel logarithmic spaces. Arch. Math. (Basel) **80**(2), 165–176 (2003)
97. Nikolova, L., Persson, L.-E., Varošanec, S.: Continuous forms of classical inequalities. Mediterr. J. Math. **13**, 3483–3497 (2016)
98. Nikolova, L., Persson, L.-E., Varošanec, S.: A new look at classical inequalities involving Banach lattice norms. J. Inequal. Appl. **2017**, 302 (2017). https://doi.org/10.1186/s13660-017-1576-8
99. Nikolova, L.I., Persson, L.-E., Varošanec, S.: On continuous versions of some norm inequalities. In: Drapaljuk, M. (ed.) Proceedings Conference "Modern Problems of the analysis: Applications in Technique and Technology", Voronezh, Sept. 2017 (Russian). Actual Directions of Scientific Research XXI Century: Theory and Practice, pp. 97–100. Springer, Berlin (2017)
100. Nikolova, L., Persson, L.-E., Samko, N.: Some new inequalities involving the Hardy operator. Math. Nach. **293**, 376–385 (2020)
101. Nikolova, L., Persson, L.-E., Varošanec, S.: Refinement of continuous forms of classical inequalities. Eurasian Math. J. **12**(2), 59–73 (2021)
102. Nikolova, L., Persson, L.-E., Varošanec, S.: Continuous refinements of some Jensen-type inequalities via strong convexity with applications. J. Inequal. Appl. **2022**, 63 (2022). https://doi.org/10.1186/s13660-022-02801-4
103. Nikolova, L., Persson, L.-E., Varošanec, S., Yimer, M.F.: Refinements of some classical inequalities via superquadracity. J. Inequal. Appl. **2022**, 86 (2022). https://doi.org/10.1186/s13660-022-02821-0
104. Oguntuase J.A., Persson, L.-E.: Refinement of Hardy's inequalities via superquadratic and subquadratic functions. J. Math. Anal. Appl. **339**(2), 1305–1312 (2008)
105. Opic, B., Kufner, A.: Hardy-type Inequalities. Pitman Research Notes in Mathematics Series 219, Longman Scientific & Technical, Harlow (1990)
106. Pachpatte, B.G.: Mathematical Inequalities. North Holland Mathematical Library, Elsevier B. V., Amsterdam (2005)
107. Pečarić, J., Beesack, P. R.: On Jessen's inequality for convex functions II. J. Math. Anal. Appl. **118**, 125–144 (1986)
108. Pečarić, J., Perić, J.: Refinements of the integral form of Jensen's and the Lah-Ribarič inequalities and applications for Cziszár divergence. J. Inequal. Appl. **2020**, 108 (2020). https://doi.org/10.1186/s13660-020-02369-x
109. Pečarić, J., Varošanec, S.: A generalization of Pólya's inequalities. In: Inequalities and Applications, World Scientific Series in Applicable Analysis, vol. 3, pp. 501–504. World Scientific Publishing Co., Inc., River Edge (1994)

110. Pečarić, J. E., Proschan, F., Tong, Y. L.: Convex Functions, Partial Ordering, and Statistical Applications. Mathematics in Science and Engineering 187, Academic Press, Inc., Boston (1992)
111. Peetre, J., Persson, L.-E.: General Beckenbach's inequality with applications. In: Function Spaces, Differential Operators and Nonlinear Analysis (Sodankylä, 1988), Pitman Research Notes in Mathematics Series, vol. 211, pp. 125–139. Longman Scientific & Technical, Harlow (1989)
112. Persson, L.-E.: Some elementary inequalities in connection with X^p spaces. In: Constructive Theory of Functions (Varna 1987), pp. 367–376. Publication House Bulgarian Academy of Sciences, Sofia (1988)
113. Persson, L.-E.: Lecture Notes. P. L. Lions Seminar (College de France, Paris) (2015)
114. Persson, L.-E., Samko, N.: What should have happened if Hardy discover this? J. Inequal. Appl. **2012**, 29 (2012). https://doi.org/10.1186/1029-242X-2012-29
115. Pólya, G., Szegö, G.: Problems and Theorems in Analysis, Vols. I and II. Classics Mathematics. Springer-Verlag, Berlin (1998)
116. Popoviciu, T.: On some inequalities. (Romanian). Gaz. Mat. Bucharest, **51**, 81–85 (1946)
117. Rooin, J.: A refinement of Jensen's inequality. JIPAM. J. Inequal. Pure Appl. Math. **6**(2), Paper No. 38 (2005), 4 p. [electronic only] http://eudml.org/doc/125164
118. Rudin, W.: Real and Complex Analysis. McGraw-Hill Book Co., New York (1987)
119. Schep, A.R.: Minkowski's integral inequality for function norms. In: Operator Theory in Function Spaces and Banach Lattices. Operator Theory: Advances and Applications, vol. 75, pp. 299–308. Birkhäuser Verlag, Basel (1995)
120. Sinnamon, G.: Refining the Hölder and Minkowski inequalities. J. Inequal. Appl. **6**(6), 633–640 (2001). http://eudml.org/doc/122104
121. Sparr, G.: Interpolation of several Banach spaces. Ann. Mat. Pura Appl. (4) **99**, 247–316 (1974)
122. Triebel, H.: Interpolation Theory, Function Spaces, Differential Operators. VEB Deutscher Verlag der Wissenschaften, Berlin (1978)
123. Varošanec, S.: A generalized Beckenbach-Dresher inequality and related results. Banach J. Math. Anal. **4**(1), 13–20 (2010)
124. Varošanec, S.: Superadditivity of functionals related to Gauss' type inequalites. Sarajevo J. Math. **10**(1), 37–45 (2014)
125. Varošanec, S., Pečarić, J.: Gauss' and related inequalities. Z. Anal. Anwendungen **14**(1), 175–183 (1995)
126. Wu, S.: A unified generalization of Aczél, Popoviciu and Bellman's inequalities. Taiwanese J. Math. **14**(4), 1635–1646 (2010)
127. Yoshikawa, A.: Sur la théorie d'espaces d'interpolation - les espaces de moyenne de plusieurs espaces de Banach (French). J. Fac. Sci. Univ. Tokyo Sect. I **16**(1969), 407–468 (1970)
128. Zhao, C.J., Cheung, W.S.: On Minkowski's inequality and its application. J. Inequal. Appl. **2011**, 71 (2011). https://doi.org/10.1186/1029-242X-2011-71

Index

B
Banach function space, 99
Banach lattice, 99
Banach lattice p-convexity, 106
Banach lattice q-concavity, 106
Beckenbach-Dresher inequality continuous form, 14
Beckenbach-Dresher inequality for functionals, 13
Beckenbach-Dresher inequality for integrals, 15
Beckenbach-Dresher inequality in Banach lattice, 113, 115
Beckenbach-Dresher inequality refinement, 73, 74
Beckenbach-Dresher inequality reverse, 116
Bellman inequality continuous form, 22
Bellman inequality discrete, 20
Bellman inequality for integrals, 21
Bellman inequality in Banach lattice, 111, 112
Bellman inequality refinement, 41, 42

C
Calderón method, 104
Complex method, 104
Convexification, 99

F
Fatou property, 100
Function strongly convex, 57
Function subquadratic, 65
Function superquadratic, 65
Functional subadditive, 89
Functional superadditive, 80

G
Gauss-Pólya inequality continuous form, 26, 28
Gauss-Pólya inequality involving derivatives, 25
Geometric mean, 3

H
Hardy-Hilbert inequality, 77
Hardy inequality for integrals, 50
Hardy inequality for superquadratic function, 76
Hardy inequality in Banach lattice, 118
Hardy inequality refinement, 51, 53
Hermite-Hadamard inequality refinement, 63
Hölder inequality continuous form, 5
Hölder inequality discrete, 2
Hölder inequality for integrals, 2, 8
Hölder inequality in Banach lattice, 100, 101
Hölder inequality refinement, 32, 33, 35, 69

I
Interpolation between families of Banach spaces, 104
Interpolation theory, 127

J
Jensen inequality for strongly convex function, 58
Jensen inequality integral, 5

Jensen inequality refinement, 45, 58, 60
Jensen inequality reverse, 43, 44
Jensen-Mercer inequality discrete, 47
Jensen-Mercer inequality for integrals, 47
Jensen-Mercer inequality refinement, 48

L
Lah-Ribarič inequality refinement, 64
Log-convex function, 128

M
Measure space, 4
Minkowski inequality continuous form, 8
Minkowski inequality discrete, 11
Minkowski inequality for integrals, 11
Minkowski inequality in Banach lattice, 104, 107
Minkowski inequality refinement, 39, 40, 70

O
Open question, 2

P
Popoviciu inequality continuous form, 17
Popoviciu inequality discrete, 15
Popoviciu inequality for integrals, 16
Popoviciu inequality in Banach lattice, 107, 108
Popoviciu inequality refinement, 36, 38

Q
Quasilinearization method, 9

R
Riesz-Thorin interpolation theorem, 3, 127

W
Weight, 5

The manufacturer's authorised representative in the EU is Springer Nature Customer Service Centre GmbH, Europaplatz 3, 69115 Heidelberg, Germany. If you have any concerns regarding our products, please contact ProductSafety@springernature.com

Printed and bound by CPI Group (UK) Ltd, Croydon, CR0 4YY

26/03/2026

02078943-0020